# Dynamic Modeling in Behavioral Ecology

MONOGRAPHS IN BEHAVIOR AND ECOLOGY

Edited by John R. Krebs and Tim Clutton-Brock

Five New World Primates: A Study in Comparative Ecology, *by John Terborgh*

Reproductive Decisions: An Economic Analysis of Gelada Baboon Social Strategies, *by R.I.M. Dunbar*

Honeybee Ecology: A Study of Adaptation in Social Life, *by Thomas D. Seeley*

Foraging Theory, *by David W. Stephens and John R. Krebs*

Helping and Communal Breeding in Birds: Ecology and Evolution, *by Jerram L. Brown*

Dynamic Modeling in Behavioral Ecology, *by Marc Mangel and Colin W. Clark*

# Dynamic Modeling in Behavioral Ecology

MARC MANGEL
COLIN W. CLARK

Princeton University Press
Princeton, New Jersey

Published by Princeton University Press,
41 William Street, Princeton, New Jersey 08540
In the United Kingdom: Princeton University Press,
Guildford, Surrey



Clothbound editions of Princeton University Press books are printed on acid-free paper, and binding materials are chosen for strength and durability. Paperbacks, although satisfactory for personal collections, are not usually suitable for library rebinding

Printed in the United States of America by Princeton University Press, Princeton, New Jersey

Designed by Laury A. Egan

Library of Congress Cataloging-in-Publication Data

Mangel, Marc.
Dynamic modeling in behavioral ecology.

(Monographs in behavior and ecology)
Bibliography: p. Includes indexes.
1. Animal behavior—Mathematical models. 2. Animal behavior—Data processing. 3. Animal ecology—Mathematical models. 4. Animal ecology—Data processing. I. Clark, Colin Whitcomb, 1931– .
II. Title. III. Series.
QL751.65.M3M29 1988 591.51′0724 88-12427
ISBN 0-691-08505-6 (alk. paper)
ISBN 0-691-08506-4 (pbk.)

To Susan and Janet

# Contents

# Preface and Acknowledgments

The purpose of this book is to explain how to construct and use dynamic optimization models in behavioral ecology.

We hope, indeed we're convinced, that our approach will clearly show that the modeling techniques we describe are well worth mastering. First, we believe that dynamic optimization models are tremendously powerful and useful for understanding many aspects of animal behavior. They contain as special cases various models that are probably already familiar—for example, life-history models. Dynamic optimization models allow for an increase in biological realism, relative to most of the models currently used in behavioral ecology. These traditional models usually pertain to a single behavioral decision, while the dynamic models cover a sequence of decisions. We will discuss many actual applications in our attempt to demonstrate the power of dynamic optimization modeling.

Equally important, dynamic optimization modeling is easy to learn. You need only a vague recollection of basic probability mathematics (which we carefully review in Chapter 1) and a desktop computer.* Of course you also need a biological problem—some behavioral phenomenon you are hoping to understand more fully.

We are indebted to the following scientists, who sent us their written comments on various portions of the book: Brian Bertram, Tom Caraco, Dan Cohen, Steve Courtney, Andy Dobson, John Endler, John Fox, Jim Gilliam, Nelson Hairston, Jr., Alasdair Houston, Dave Levy, Guy Martel, Craig Packer, Bernie Roitberg, Earl Werner, and Dave Winkler. We also owe a debt of gratitude to the students in our courses taught at the University of British Columbia, University of California (Davis), Cornell University, The Hebrew University of Jerusalem, and Simon Fraser University.

Our research has been financially assisted by NSF Grant BSR 86-1073 and NSERC Grant A-3990. M.M. acknowledges the receipt of a Guggenheim fellowship and a Fulbright fellowship. C.C. is grateful to the University of California (Davis), and to Cornell University for generously hosting him while the book was in progress.

* For the mathematically literate reader we remark here that the dynamic optimization models to be discussed in this book are in fact stochastic dynamic programming models. This explains both the need for probability theory, and the use of the computer.

Janet Clark TEX’d the numerous drafts of the book.

A special acknowledgment is reserved for Saturna Island, a peaceful place of peregrines and orcas.

# Dynamic Modeling in Behavioral Ecology

# Introduction

## Optimization Modeling in Behavioral Ecology

If by "model" we mean simply a description of nature, then optimization models have played a central role in evolutionary biology ever since the publication of Charles Darwin's *The Origin of Species.* For if we adopt the hypothesis that natural selection leads (within limits) to the "survival of the fittest," then we can hope to understand the observed morphology, physiology, and behavior of living organisms on the basis of optimization principles. Darwin's hypothesis provides a unifying principle in biology which is every bit as powerful as Hamilton's principle of least action in physics.

Early optimization models in biology were largely descriptive. The advantages of a more rigorous approach based on quantitative, mathematical optimization models have recently been recognized. In behavioral ecology, for example, the development of a class of models subsumed under the title "optimal foraging theory" has generated new hypotheses and suggested new experiments concerning the influence of environmental factors on the behavior of animals (Krebs and Davies 1984, Stephens and Krebs 1986). Similarly, game theory (a branch of optimization theory) has led to new insights into behavioral interactions between animals (Maynard Smith 1982).

In spite of, perhaps even as a result of, the undeniable success of optimization modeling in biology, two general classes of criticism of the so-called "adaptationist paradigm" have arisen. Because we believe that these criticisms are cogent, and also because we believe that the modeling methodology presented in this book goes some distance (but certainly not all the way) towards meeting both criticisms, we wish briefly to outline their main points.

The first criticism, exemplified by Gould and Lewontin (1979), finds fault with models which suppose that each and every separate "trait" must have evolved optimally in terms of fitness. The genetic mechanism, however, does not usually allow traits (however they may be defined) to evolve independently. Also, except for traits that affect fitness independently, optimizing one trait will not be consistent with optimizing a different trait. One must therefore consider "tradeoffs," and not merely between pairs of traits. The genetic mechanism also sets constraints: existing traits can only have evolved from pre-existing traits. While Gould and Lewontin list several other criticisms of the adaptationist paradigm, their overall conclusion (stated somewhat simplistically) is that optimization analysis is indeed a valid and powerful paradigm in evolutionary biology, but organisms must be treated as integrated wholes, constrained by their phyletic heritage.

It is our view that, valuable as it may have been for the initial development of a new field of research, much of the early modeling work in behavioral ecology is vulnerable to the foregoing criticism. These models were almost always concerned with a single decision—choice of patch, time to remain in a patch, choice of food items, optimal clutch size, and so on. The models were static, and either deterministic or based on long-term averages of random processes. In other words, each model treated a single behavioral choice, which was isolated from other behavioral choices that might be available simultaneously or subsequently.

Behavioral ecologists have been fully aware of the limitations of treating different behavioral "traits" independently, but have been largely stumped by the need to discover a "common currency" for different kinds of decisions.

The problem of treating an animal's behavior as an integrated whole has been discussed in general terms by the behavioral ecology group at Oxford University (see McFarland and Houston 1981, McFarland 1982), who clearly recognized the need for a *state-space approach* to the modeling of behavior and ontogeny. The present book is very much consistent with this line of thought: we agree that a state-space approach is essential. By adopting the relatively elementary methodology of discrete-time, stochastic dynamic programming, rather than the more esoteric techniques of continuous-time optimal control theory, and relying on computer rather than analytic solution methods, we believe that the modeling approach described herein is both simple and implementable in practical terms.

In particular, a principal thrust of the present work is to provide a more unified approach to the modeling of animal behavior from the evolutionary standpoint. Our approach allows for an arbitrary number of types of behavior to be considered both simultaneously and sequentially; the common currency is fitness defined as expected lifetime reproduction. Furthermore, physiological and environmental constraints can easily be incorporated into our framework. Indeed they can hardly be left out.

The second type of criticism (see, for example, Oster and Wilson 1978, Ch. 8) goes roughly like this: any behavioral model which is simple enough to be operational is necessarily too simple to be biologically realistic. Obversely, any biologically realistic model in behavioral ecology will be too complex to be operational (and, we might add, to be mathematically tractable). Again, while admitting the validity of this criticism, Oster and Wilson nevertheless conclude that "the construction and testing of [optimization] models is a potentially powerful technique analogous to blind experimentation conducted in the laboratory."

We wholeheartedly concur with all this. But we also hope that the modeling approach described in this book will help to narrow the gap between what is realistic and what is operational in behavioral modeling. In a sense, the gap now seems largely technological. Faster computers and better data sets will slowly allow the gap to be further narrowed.

Perhaps the final limitation to modeling will be human understanding. If our models become as complex as nature itself they will be as difficult to understand as nature is. This is not a joke: the point of modeling is to increase our understanding of nature, not merely to reproduce it in the computer.

## How Do Organisms Optimize?

Can organisms really "solve" complicated dynamic optimization problems? Clearly animals do not rationalize about risk, or compute the reproductive potential corresponding to alternative behavioral strategies. In ways that we are barely beginning to comprehend, behavior, even flexible, responsive behavior, is "hardwired" into the genetic system. Oftimes behavior is so finely tuned that we poor humans are at a loss to explain the how or the why of it.

In this book we are going to accept the following two hypotheses without further discussion:

(1) Any significant adaptive advantage that is phyletically feasible will tend to be selected;

(2) Organisms do have some way of getting near to optimal solutions of behavioral problems in situations that they normally encounter.

What is one to conclude if, having constructed an optimization model, one discovers that the animals in question fail to behave according to predictions? Perhaps evolution has not yet achieved the *optimum optimorum*; the population may be stuck at a local optimum, or environmental changes may have shifted the optimum. But proffering such an explanation is tantamount to an admission of defeat for the modeling project.

We believe that in many cases the greater truth lies in recognizing that the model is inadequate. Later in the book (Chapters 3–7) we give numerous examples of how the dynamic approach to behavioral modeling has led to new and different predictions which are more in accord with observations. Perhaps evolution is more adept at locating optima than we are in discovering in what ways a certain behavior is adaptive.

Another possible reason for the failure of optimization models to explain observations arises under laboratory conditions, where animals are often exposed to behavioral problems remote from those they would be likely to encounter in nature. Can one really expect animals to overcome their genetic limitations, and solve unfamiliar behavioral problems optimally according to our concept of optimality? This rather trivial consideration is overlooked with surprising frequency.

## How to Use This Book

The nine chapters of this book are divided into three Parts. Part I (Chapters 1–2) contains essential instructional material for anyone wishing to take up the art of optimization modeling of behavior.

Chapter 1 gives the absolute minimum of background in probability. This material should be familiar, but perhaps not second nature, to most readers. Chapter 2 introduces the techniques of dynamic modeling by means of a very simple basic paradigm, the problem of optimal patch selection when different patches have different levels of resource abundance and different risks of predation. Modeling this situation has until recently been considered quite beyond the scope of mathematical treatment. We hope that Chapter 2 will demonstrate how simple it is to construct such models. Chapters 1 and 2 contain several appendices, which can be skipped at a first reading. However, some of these appendices are required for the applications in Part II; we will let you know when you need a particular appendix. Part I ends with an addendum, which is a brief guide to the writing of computer programs for beginners. You will need to read this only if you have never written a program before.

The five chapters of Part II are devoted to actual applications. They cover a fairly wide range of taxonomy and of behavioral observations. These applications are presented as illustrations of the scope of the dynamic modeling approach. They are by no means exhaustive treatments of their particular subjects; in fact most of these subjects would be worthy of an entire book.

In Part III we present some perspectives on the dynamic modeling approach discussed in this book. Chapter 8 provides a general discussion of the techniques involved in formulating and solving a dynamic optimization model of behavior. A brief description of various alternative modeling techniques, and their relationship to the dynamic approach used in this book, also appears in Chapter 8. Chapter 9 discusses two extensions of the basic modeling approach, first to include informational aspects, and second to include game-theoretic aspects of behavior. Both of these subjects greatly increase the degree of computational difficulty likely to be encountered in solving a dynamic optimization model.

Finally, in the Epilogue we discuss some general principles that we believe pertain to the dynamic modeling of behavior.

For serious readers of this book we recommend first a careful perusal of Chapters 1 and 2. *Take nothing on faith.* Make sure that you fully understand all notation. Every equation should be read and understood (*not* memorized!) completely. If in doubt, write out the details yourself, as if you were preparing to lecture on the subject. We have made every effort to keep the mathematics as simple as possible,* and nothing in these chapters is superfluous to the applications in Part II.

* Our professional philosophy is "learn new things when you need them."

# I Fundamentals

*Part I presents the necessary mathematical background for formulating dynamic models in behavioral ecology. In Chapter 1 we review some fundamental concepts and notations from elementary probability theory. The material in the main part of this chapter will be used continually throughout the book. The appendices to Chapter 1 can be skipped over at first reading, but each eventually comes into play in the applications of Part II.*

*Chapter 2 is essential reading, as it introduces the modeling technique used throughout the book, but in a deliberately abstract and simple setting, that of optimal patch selection when different patches have different levels of food abundance and different risks of predation. The appendices to Chapter 2 are really an integral part of the chapter, showing how the basic model can be extended to study many additional aspects of the patch selection problem. Readers anxious to see "real-world" applications may wish to skim over these appendices quickly and pass on to one or more chapters in Part II. However, most of the methods described in these appendices will be used in Part II also.*

# 1 Basic Probability

In this chapter we give the necessary minimum background in probability required for dynamic stochastic modeling. We anticipate that the reader has encountered most of this material elsewhere, but we do not assume a working familiarity with it.

The main text of this chapter contains all the mathematics needed to read Chapter 2, where we introduce the basic modeling approach. The appendices of this chapter contain additional material that is used occasionally in the later parts of the book.

## 1.1 Notation

In probability theory we are concerned with chance occurrences, called *events*. The probability that a certain event $A$ occurs will be denoted by

$$\Pr(A)$$

(read as "the probability of $A$," or "the probability that $A$ occurs"). Thus we will encounter expressions like Pr (animal survives $T$ periods) and $\Pr(X(T) = 0)$.

### Frequency Interpretation

Probability has an intuitive "frequency" interpretation as follows. Imagine an "experiment" in which a certain event $A$ is one of various possible outcomes; which outcome actually occurs in a replication of the experiment depends on chance. Then $\Pr(A)$ represents the average proportion of occurrences of the event $A$ in a long series of $N$ repetitions of the identical experiment:

$$\Pr(A) = \lim_{N\to\infty} \frac{\text{number of occurrences of } A \text{ in } N \text{ repetitions}}{N}.$$

It follows that

$$0 \leq \Pr(A) \leq 1 \tag{1.1}$$

$$\Pr(\text{no event}) = 0; \ \Pr(\text{some event}) = 1 \tag{1.2}$$

and also that

$$\Pr(A \text{ or } B) = \Pr(A) + \Pr(B) \tag{1.3}$$

if $A$ and $B$ are mutually exclusive events. (The reader should pause to verify that Eqs.(1.1)–(1.3) do indeed follow logically from the above frequency interpretation.)

### Set-theoretic Notation

In working with probability, it is convenient to use the abstract, set-theoretic notation encountered in modern texts. In the abstract setting, probabilities are usually defined for "sets" $A$; that is, events are thought of as subsets $A, B, \ldots$ of some universal set $S$ called the *sample space* of the experiment. The elements of $S$ may be thought of as basic events. For example, suppose that $S$ consists of the set of all possible 13-card bridge hands; $A$ is the event that "the hand contains the Ace of Spades;" $B$ is the event that "the hand contains exactly 4 cards in the suit Spades." Then, assuming that all hands are equally likely, we have $\Pr(A) = 1/4$, and $\Pr(B)$ could be computed with a lot of work (by enumerating all hands containing four spades).

Combinations of events therefore correspond to combinations of sets. In particular we have

$$A \text{ or } B = A \cup B$$
$$A \text{ and } B = A \cap B$$

where $A \cup B$ (read "$A$ union $B$") is the set of all elements of $S$ which belong to $A$ or to $B$ (or both), and $A \cap B$ (read "$A$ intersect $B$") is the set of all elements of $S$ which belong to both $A$ and $B$. These set operations, as they are called, can be pictured in terms of Venn diagrams, as shown in Figure 1. (What would $A \cup B$ and $A \cap B$ consist of for the bridge-hand example?)

The empty set (i.e. the set of no events) is denoted by $\phi$. Eqs. (1.1)–(1.3) can now be rewritten in the form

$$0 \leq \Pr(A) \leq 1 \tag{1.4}$$

$$\Pr(\phi) = 0; \quad \Pr(S) = 1 \tag{1.5}$$

$$\Pr(A \cup B) = \Pr(A) + \Pr(B) \qquad \text{if } A \cap B = \phi. \tag{1.6}$$

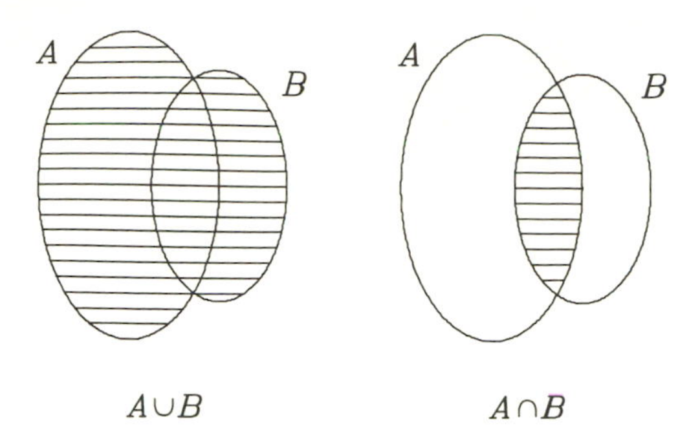


Figure 1.1 Venn diagrams for set union and intersection.

Eq. (1.6) extends to any number of sets:

$$\Pr(A \cup B \cup C \cup \cdots) = \Pr(A) + \Pr(B) + \Pr(C) + \cdots$$

if $A, B, C, \ldots$ are mutually disjoint sets.

### Conditional Probability

Perhaps the most important idea from probability theory for the purpose of this book is the concept of conditional probability. If $A$ and $B$ are two events, then the ***conditional probability of A given B*** is defined as

$$\Pr(A \mid B) = \frac{\Pr(A \text{ and } B)}{\Pr(B)} = \frac{\Pr(A \cap B)}{\Pr(B)} \tag{1.7}$$

where the left-hand side is read as "Probability of $A$ given $B$." The frequency interpretation of conditional probability is as follows. An experiment is repeated many times. Then $\Pr(A \mid B)$ is the proportion of times that $A$ occurs in those repetitions in which $B$ occurs.

For example, suppose two fair coins are tossed, and let $A$ be the event "two heads," and $B$ the event "at least one head." The sample space of this experiment consists of the four equally likely outcomes

$$HH, HT, TH, TT.$$

Note that $A = HH$ occurs 1/3 of the time that $B = (HH, HT, TH)$ occurs. Eq. (1.7) gives the same result:

$$\Pr(A \mid B) = \frac{\Pr(A \text{ and } B)}{\Pr(B)} = \frac{1/4}{3/4} = \frac{1}{3}.$$

It is worthwhile noting that a conditional probability is itself another probability, in the sense that (for fixed $B$), $\Pr(A \mid B)$ satisfies the three axioms (1.4)–(1.6).*

Two events $A$ and $B$ are said to be ***independent*** if

$$\Pr(A \mid B) = \Pr(A). \tag{1.8}$$

This says that the knowledge that $B$ has occurred does not affect the probability that $A$ will occur. Hence the name "independent." Note that if $A$ and $B$ are independent then (1.8) implies, by (1.7),

$$\Pr(A \cap B) = \Pr(A)\Pr(B), \tag{1.9}$$

and hence also that

$$\Pr(B \mid A) = \frac{\Pr(B \cap A)}{\Pr(A)} = \Pr(B). \tag{1.10}$$

That is, the probability of $B$ occurring is independent of whether $A$ occurs or not. Any one of Eqs. (1.8)–(1.10) could be used as the definition of independence of $A$ and $B$.

## Law of Total Probability

The notion of conditional probability is often extremely useful for computations because of the following.

**Law of Total Probability**: If the sets $B_i$ $(i = 1, 2, \ldots)$ are mutually disjoint and exhaustive,**then for any event $A$,

$$\Pr(A) = \Sigma_i \Pr(A \mid B_i)\Pr(B_i). \tag{1.11}$$

The proof is quite simple: because the $B_i$ are mutually disjoint and exhaustive (see Figure 1.2), we have

$$\begin{aligned}\Pr(A) &= \Sigma_i \Pr(A \cap B_i)\\ &= \Sigma_i \Pr(A \mid B_i)\Pr(B_i) \qquad \text{by (1.7).}\end{aligned}$$

* The reader may wish to prove this for himself or herself; to prove (1.6), use the set logic identity $(A_1 \cup A_2) \cap B = (A_1 \cap B) \cup (A_2 \cap B)$, which can be verified by means of a Venn diagram.

** In set notation, the sets $B_i$ are mutually disjoint if $B_i \cap B_j = \phi$ for $i \neq j$; they are exhaustive if $B_1 \cup B_2 \cup \cdots = S$.

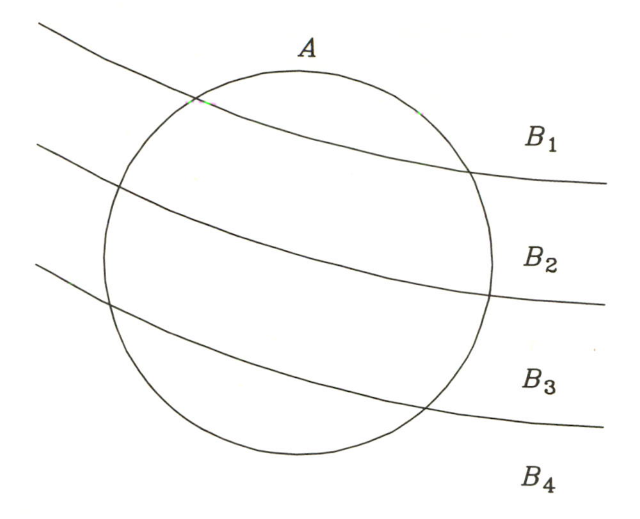


Figure 1.2 The law of total probability.

Not only is this a rigorous proof, it also explains the Law of Total Probability intuitively (see the figure): the probability of $A$ is the sum of the probabilities of ($A$ and $B_i$), which can be written as $\Pr(A \mid B_i)\Pr(B_i)$. In practice it is sometimes easy to figure out what these latter probabilities are. We will use this method repeatedly in this book.

## 1.2 Discrete Random Variables and Distributions

In setting up probabilistic models, one must be adept at handling the elementary formulas of the preceding section. Equally important is the understanding of random variables, distributions, and expectation.

### Random Variables

The term *random variable* is used heuristically to refer to a numeric variable whose value is determined by chance. Annual rainfall in Chicago would be one among countless examples. Another would be the sum of the two top faces when two dice are thrown.

A random variable that can assume only discrete values $x_1$, $x_2, \ldots$ is called a *discrete* random variable, whereas a random variable capable of assuming a continuum of values is called a *continuous* random variable. An example of a discrete random variable would be the number of offspring produced by some animal. The body weight of an animal would be an example of a continuous random variable. (However, in applications body weight can often be represented adequately by a discrete random variable, as we usually do in this book.)

For a discrete random variable $X$, the probabilities

$$p_k = \Pr(X = x_k) \tag{1.12}$$

are said to define the *probability distribution* of $X$. For example, the probabilities for the sum of two dice are:

| $x_k$ | 2 | 3 | 4 | 5 | 6 | 7 | 8 | 9 | 10 | 11 | 12 |
|---|---|---|---|---|---|---|---|---|---|---|---|
| $p_k$ | $\frac{1}{36}$ | $\frac{2}{36}$ | $\frac{3}{36}$ | $\frac{4}{36}$ | $\frac{5}{36}$ | $\frac{6}{36}$ | $\frac{5}{36}$ | $\frac{4}{36}$ | $\frac{3}{36}$ | $\frac{2}{36}$ | $\frac{1}{36}$ |

(Why? Let $(i, j)$ denote the values that appear on die 1 and die 2 respectively. Then

$$\Pr(X = 2) = \Pr(1,1) = \frac{1}{6} \times \frac{1}{6} = \frac{1}{36}$$
$$\Pr(X = 3) = \Pr(1,2) + \Pr(2,1) = \frac{1}{6} \times \frac{1}{6} + \frac{1}{6} \times \frac{1}{6} = \frac{2}{36}$$

and so on.)

The sum of all the probabilities must always equal one:

$$\sum_k p_k = 1.$$

## Expectation

The *expectation* of a discrete random variable $X$ is defined as

$$E\{X\} = \Sigma_k p_k x_k. \tag{1.13}$$

The expectation of $X$ is also called the mean or average of $X$. When the frequency interpretation is adopted, we have (for large $N$)

$$\begin{aligned} E\{X\} &= \sum_k p_k x_k \\ &= \sum_k \frac{(\text{number of times } X = x_k)}{N} x_k \end{aligned} \tag{1.14}$$

and this is exactly the usual average of $X$.

More generally, for any function $g(x)$ we define

$$E\{g(X)\} = \Sigma_k p_k g(x_k). \tag{1.15}$$

We use the notation

$$\mu_X = E\{X\} \tag{1.16}$$

and

$$\sigma_X^2 = E\{(X - \mu_X)^2\} \tag{1.17}$$

for the mean and *variance* of $X$, respectively. Eq. (1.17) tells us that the variance of $X$ is equal to the average of the square of the difference between $X$ and its mean. Thus the variance is a measure of the "spread" of $X$ away from its mean. A random variable with high variance "does a lot of varying." (What would zero variance imply?)

A dimensionless description of the amount of variation exhibited by a random variable $X$ is the *coefficient of variation*, defined by

$$CV_X = \frac{\sigma_X}{\mu_X}.$$

For example, the statement that $CV_X = 50\%$ carries the rough mental interpretation that the random variable $X$ "typically varies $\pm 50\%$ from its mean value."

It follows from the definition (1.15) that expectation has the *linearity* property:

$$E\{c_1 g_1(X) + c_2 g_2(X)\} = c_1 E\{g_1(X)\} + c_2 E\{g_2(X)\}. \tag{1.18}$$

Using this, we have

$$\begin{aligned}
\sigma_X^2 &= E\{(X - \mu_X)^2\} \\
&= E\{X^2 - 2\mu_X X + \mu_X^2\} \\
&= E\{X^2\} - 2\mu_X E\{X\} + \mu_X^2 E\{1\} \\
&= E\{X^2\} - 2\mu_X^2 + \mu_X^2 \qquad \text{since } E\{X\} = \mu_X \text{ and } E\{1\} = 1 \\
&= E\{X^2\} - \mu_X^2.
\end{aligned}$$

Hence

$$\sigma_X^2 = E\{X^2\} - \mu_X^2. \tag{1.19}$$

This formula is often used in calculating variances.

## 1.3 Conditional Expectation

Closely related to the notion of conditional probability is the concept of conditional expectation. Let $X, Y$ be two random variables. Then, according to the definition (1.7) of conditional probability, we have

$$\Pr(X = x_k \mid Y = y_j) = \frac{\Pr(X = x_k \text{ and } Y = y_j)}{\Pr(Y = y_j)} \tag{1.20}$$

and, for any given value of $y_j$, this is also a probability distribution called the *conditional distribution of $X$ given $Y = y_j$*. The *conditional expectation of a function $g(X)$ given $Y = y_j$* is then defined as the expectation with respect to this conditional distribution:

$$E\{g(X)|Y = y_j\} = \Sigma_k g(x_k) \Pr(X = x_k|Y = y_j). \tag{1.21}$$

We have immediately the important formula:

**Law of Total Expectation**:

$$E\{g(X)\} = \Sigma_j E\{g(X)|Y = y_j\} \Pr(Y = y_j). \tag{1.22}$$

This is easily derived from the Law of Total Probability (1.11), as follows:

$$\begin{aligned} E\{g(X)\} &= \Sigma_k g(x_k) p(x_k) \\ &= \Sigma_k g(x_k) \Sigma_j \Pr(X = x_k|Y = y_j) \Pr(Y = y_j) \quad \text{by}(1.11) \\ &= \Sigma_j (\Sigma_k g(x_k) \Pr(X = x_k|Y = y_j)) \Pr(Y = y_j) \quad \text{(why?)} \\ &= \Sigma_j E\{g(X)|Y = y_j\} \Pr(Y = y_j) \qquad \text{by}(1.21). \end{aligned}$$

Equation (1.22) has exactly the same intuitive content as the Law of Total Probability, Eq. (1.11)—namely, the expectation of $g(X)$ is the sum of expectations conditioned on values of $Y$ times the probabilities of those values.

We will see in Chapter 2 that the Law of Total Expectation is fundamental for the derivation of the basic equations used throughout the rest of the book.

The reader now has enough basic mathematics to read Chapter 2 and most of the applications chapters. The four appendices here provide additional mathematical background for stochastic modeling of behavior. Most of this material will be used occasionally in later chapters.

## Appendix 1.1
## The Poisson Process

We now consider a random process taking place over time, in which events of a certain type can occur at various points in time. Examples might include: telephone calls arriving at a switchboard, alpha particles emitted from a radioactive source, or discoveries of food items by a forager.

Such processes can be modeled by a mathematical construction called the *Poisson* process, which is defined by the conditions

$$\left.\begin{array}{l}\Pr\ (\text{one event in } dt) \approx \lambda\, dt \\ \Pr\ (\text{no event in } dt) \approx 1 - \lambda\, dt \\ \Pr\ (\text{two or more events in } dt) \approx 0\end{array}\right\} \qquad (1.23)$$

for small time intervals $dt$. The symbol $\approx$ is read "is approximately equal to"; the strict mathematical meaning of Eq. (1.23) must be expressed in terms of limits:

$$\frac{\Pr\ (\text{one event in } dt)}{dt} \to \lambda \text{ as } dt \to 0.$$

Some books use the "little-oh" notation:

$$\left.\begin{array}{l}\Pr\ (\text{one event in } dt) = \lambda\, dt + o(dt) \\ \Pr\ (\text{no event in } dt) = 1 - \lambda\, dt + o(dt) \\ \Pr\ (\text{two or more events in } dt) = o(dt)\end{array}\right\} \qquad (1.24)$$

where the symbol $o(dt)$, read as "little-oh of dt," signifies unspecified terms which approach zero when divided by $dt$, as $dt \to 0$. For example, $(dt)^2 = o(dt)$. Such terms are said to be *of smaller order* than $dt$.

The parameter $\lambda$ in Eq. (1.23) is a constant, called the *rate parameter* of the Poisson process. We will see that $\lambda$ represents the average rate of occurrences of the event.

For a given Poisson process, let $X_t$ denote the number of events that occur in a fixed time interval $[0,t]$. Then $X_t$ is a random variable with distribution

$$p_k(t) = \Pr(X_t = k) \qquad k = 0, 1, 2, \ldots$$

In principle $X_t$ can take any integer value greater than or equal to zero. Some values are more likely than others, so we want to find the $p_k(t)$. This is done as follows. First, since $p_0(t) = \Pr$ (no events in $[0,t]$) we have

$$\begin{aligned} p_0(t+dt) &= \Pr \text{ (no events in } [0,t+dt]) \\ &= \Pr \text{ (no events in } [0,t]) \cdot \Pr \text{ (no events in } [t,t+dt]) \\ &\approx p_0(t) \cdot (1 - \lambda\, dt) \quad \text{by (1.23)} \end{aligned}$$

so that

$$\frac{p_0(t+dt) - p_0(t)}{dt} \approx -\lambda p_0(t).$$

Now let $dt \to 0$:

$$\frac{dp_0}{dt} = -\lambda p_0.$$

Noting that $p_0(0) = 1$ (i.e. the probability of no events up to time zero is one!), we see that

$$p_0(t) = e^{-\lambda t}. \tag{1.25}$$

This result is intuitively appealing: the probability that no event occurs between 0 and $t$ decreases at an exponential rate (determined by $\lambda$) as $t \to \infty$.

The same method can be used to find $p_1(t)$:

$$\begin{aligned} p_1(t+dt) &= \Pr \text{ (one event in } [0,t+dt]) \\ &= \Pr \text{ (no events in } [0,t]) \cdot \Pr \text{ (one event in } dt) \\ &\qquad + \Pr \text{ (one event in } [0,t]) \cdot \Pr \text{ (no events in } dt) \\ &\approx p_0(t)\lambda\, dt + p_1(t)(1 - \lambda\, dt). \end{aligned}$$

Simplifying as before, we obtain the differential equation

$$\frac{dp_1}{dt} + \lambda p_1 = \lambda p_0(t) = \lambda e^{-\lambda t}.$$

Also $p_1(0) = 0$ (why?). This differential equation can be solved by methods taught in a first course in differential equations. The solution is in fact

$$p_1(t) = \lambda t e^{-\lambda t} \tag{1.26}$$

which can also be verified by direct substitution into the differential equation.

The same method can be extended indefinitely. In general we have

$$\begin{aligned} p_{n+1}(t+dt) &= \Pr\ (n+1 \text{ events in } [0, t+dt]) \\ &= \Pr\ (n \text{ events in } [0,t]) \cdot \Pr\ (\text{one event in } [t, t+dt]) \\ &\quad + \Pr\ (n+1 \text{ events in } [0,t]) \cdot \Pr\ (\text{no events in } [t, t+dt]) \\ &\approx p_n(t) \cdot \lambda\, dt + p_{n+1}(t)(1 - \lambda\, dt). \end{aligned}$$

As before, this leads to a differential equation:

$$\frac{dp_{n+1}}{dt} + \lambda p_{n+1} = \lambda p_n(t). \tag{1.27}$$

Hence, once $p_n(t)$ is known, $p_{n+1}(t)$ can be obtained by solving this differential equation. For example, when $n = 2$, we have

$$\frac{dp_2}{dt} + \lambda p_2 = \lambda p_1(t) = \lambda^2 t e^{-\lambda t}.$$

Again, this can be solved to obtain

$$p_2(t) = \frac{(\lambda t)^2}{2} e^{-\lambda t}.$$

At this stage, one might be tempted to guess the form of the general case. The correct guess turns out to be

$$p_n(t) = \frac{(\lambda t)^n}{n!} e^{-\lambda t}. \tag{1.28}$$

(Recall that $n!$ is read as "$n$ factorial," and is defined by $n! = 1 \cdot 2 \cdot 3 \cdots n$ and (special case) $0! = 1$.) We will not attempt to prove this in general, but the reader may wish to check that (1.27) is indeed satisfied if $p_n(t)$ is given by (1.28).

From (1.28) we see that*

$$\sum_{n=0}^{\infty} p_n(t) = e^{-\lambda t} \sum_{n=0}^{\infty} \frac{(\lambda t)^n}{n!} = 1. \tag{1.29}$$

This provides a check that the probabilities $p_n(t)$ given by (1.28) sum to 1, as they must.

* Recall that $\Sigma_0^\infty x^n/n! = e^x$ from elementary calculus.

### COMPUTATION OF THE POISSON PROCESS

Observe that the values of $p_n(t)$ given by Eq. (1.28) depend on the product $\lambda t$, and not on $\lambda$ and $t$ respectively. We now suggest that the reader use a calculator or simple microcomputer program to calculate $p_n(t)$ for $\lambda t = 5$ (say), and $n = 0, 1, 2, \ldots$. Which $n$ gives the largest value? Can you see why?

There is an easy way and a hard way to do this calculation. The hard way is simply to calculate each $p_n(t)$, for $n = 0, 1, 2, \ldots$, separately. The easy way is to notice that each $p_n(t)$ is closely related to the previous one. From (1.28) we see that

$$p_n(t) = \frac{\lambda t}{n} p_{n-1}(t). \tag{1.30}$$

Hence, starting with $p_0(t) = e^{-\lambda t}$ one can quickly calculate $p_1(t)$, $p_2(t)$ and so on. This algorithm* is very easy to program on the computer. We suggest the reader experiment with a few calculations; similar methods are very often useful.

The factor $\lambda t/n$ appearing in (1.30) is greater than one if $n < \lambda t$, and less than one if $n > \lambda t$. This means that

$$p_n(t) > p_{n-1}(t) \quad \text{if} \quad n < \lambda t \quad \text{and vice versa.}$$

Hence $p_n(t)$ increases up to $n = \lambda t$ and then decreases—i.e. the probability $p_n(t)$ is largest for $n = \lambda t$. (Can you determine the largest term if $\lambda t$ is not an integer?)

An example of the Poisson probability distribution is shown in Figure 1.3.

### MEAN AND VARIANCE

The mean of the Poisson distribution is

$$\begin{aligned} \mu_{X_t} = E\{X_t\} &= \sum_{n=0}^{\infty} n p_n(t) \\ &= e^{-\lambda t} \sum_{n=0}^{\infty} n \frac{(\lambda t)^n}{n!}. \end{aligned} \tag{1.31}$$

* An *algorithm* is just a method for computing something. Algorithms are *coded* into computer programs. If you are a total beginner at computer programming, the Addendum to Part I may be helpful.

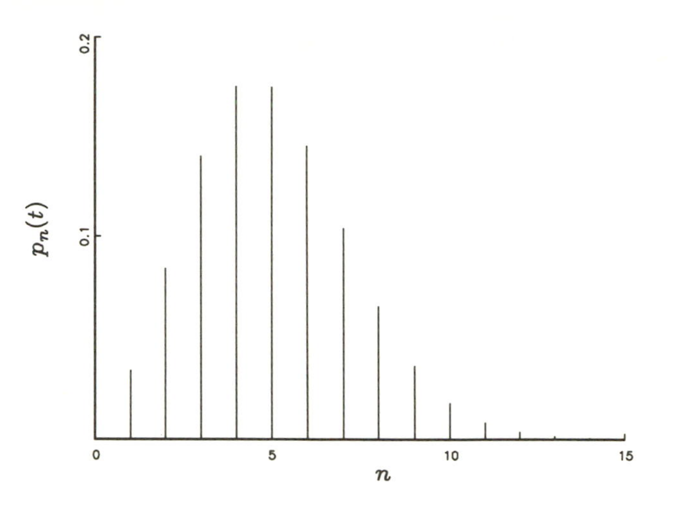


Figure 1.3 Poisson probability distribution $p_n(t)$, for $\lambda t = 5.0$.

The following method of evaluating the infinite sum in the last line may seem like a mathematical "trick," but in fact the method of evaluating sums by differentiation is often extremely useful.* Namely, we combine the two formulas from elementary calculus

$$\frac{d}{dx}x^n = nx^{n-1} \quad \text{and} \quad \sum_{n=0}^{\infty} \frac{x^n}{n!} = e^x$$

as follows:

$$\begin{aligned}
\sum_{n=0}^{\infty} n\frac{x^n}{n!} &= \sum_{0}^{\infty} x\frac{nx^{n-1}}{n!} \\
&= x\sum_{0}^{\infty} \frac{d}{dx}\left(\frac{x^n}{n!}\right) \\
&= x\frac{d}{dx}\sum_{0}^{\infty} \frac{x^n}{n!} \\
&= x\frac{d}{dx}e^x = xe^x.
\end{aligned}$$

* The late Nobel laureate physicist Richard Feynman was perhaps the leading advocate of such methods.

Substituting $x = \lambda t$ in Eq. (1.31) above, we find that

$$\mu_{X_t} = E\{X_t\} = e^{-\lambda t} \cdot \lambda t e^{\lambda t} = \lambda t. \tag{1.32}$$

This result agrees with intuition: if $\lambda$ is the average rate of occurrence of events, then the average number of events in time $t$ is $\lambda t$. As we showed above, the most likely number of events to occur in time $t$ is also equal to $\lambda t$ (or the greatest integer in $\lambda t$)—see Figure 1.3.

An alternative method for computing the expectation works directly with the summation:

$$\begin{aligned} E\{X_t\} &= \sum_{n=0}^{\infty} n p_n(t) = \sum_{n=0}^{\infty} \frac{n e^{-\lambda t} (\lambda t)^n}{n!} \\ &= e^{-\lambda t} \sum_{n=0}^{\infty} n \frac{(\lambda t)^n}{n!}. \end{aligned} \tag{1.33}$$

The first term in the sum is identically 0 (recall $0! = 1$ by definition) so

$$E\{X_t\} = e^{-\lambda t} \sum_{n=1}^{\infty} (\lambda t)^n / (n-1)! \tag{1.34}$$

Now set $j = n - 1$; when $n = 1$, $j = 0$ so that

$$\begin{aligned} E\{X_t\} &= e^{-\lambda t} \sum_{j=0}^{\infty} (\lambda t)^{j+1} / j! \\ &= e^{-\lambda t} \cdot \lambda t \cdot \sum_{j=0}^{\infty} (\lambda t)^j / j! \\ &= e^{-\lambda t} \lambda t \cdot e^{\lambda t} \\ &= \lambda t \end{aligned} \tag{1.35}$$

as before. Either method can be used to calculate $E\{X_t^2\}$. For example,

$$n^2 x^n = x \frac{d}{dx}(n x^n).$$

Therefore

$$\sum_0^\infty n^2 \frac{x^n}{n!} = x \frac{d}{dx}\Big(\sum_0^\infty n \frac{x^n}{n!}\Big)$$
$$= x \frac{d}{dx}(xe^x) = xe^x + x^2 e^x.$$

Hence

$$E\{X_t^2\} = \sum_0^\infty n^2 \frac{(\lambda t)^n}{n!} e^{-\lambda t}$$
$$= e^{-\lambda t}\Big(\lambda t e^{\lambda t} + (\lambda t)^2 e^{\lambda t}\Big)$$
$$= \lambda t + (\lambda t)^2.$$

From this we obtain

$$\sigma_{X_t}^2 = E\{X_t^2\} - \mu_{X_t}^2 = \lambda t + (\lambda t)^2 - (\lambda t)^2$$

or

$$\sigma_{X_t}^2 = \lambda t. \tag{1.36}$$

The Poisson distribution has the unusual property that the mean and the variance are the same. Both increase proportionally to the length of time that the process is observed. The coefficient of variation

$$CV_{X_t} = \frac{\sigma_{X_t}}{\mu_{X_t}} = \frac{1}{\sqrt{\lambda t}}$$

decreases as $t$ increases.

## Appendix 1.2
## Continuous Random Variables

We now consider a random variable $X$ that can take on a continuum of values instead of only discrete values. To study this type of random variable, we introduce the *cumulative distribution function*, defined as*

$$F_X(x) = \Pr\{X \leq x\} \qquad (-\infty < x < \infty). \tag{1.37}$$

* The symbol $F_X(x)$ is read as "capital eff sub capital ex of small ex."

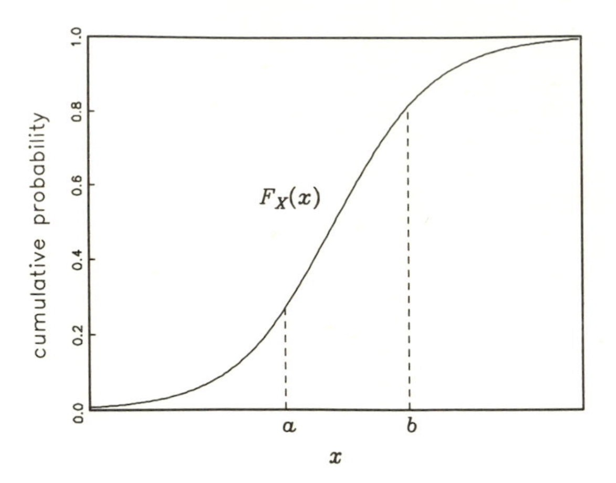


Figure 1.4 The cumulative distribution $F_X(x)$ of a continuous random variable $X$.

Figure 1.4 shows a typical example of a cumulative distribution function. From the equation (see Figure 1.4)

$$\Pr(X \leq b) = \Pr(X \leq a) + \Pr(a < X \leq b) \quad \text{(for } a < b\text{)}$$

we conclude that

$$F(b) - F(a) = \Pr(a \leq X \leq b). \tag{1.38}$$

We now define the *density* $f_X(x)$ of a continuous random variable as the derivative of the cumulative distribution function:

$$f_X(x) = \frac{d}{dx} F_X(x). \tag{1.39}$$

The intuitive content of the density follows from Taylor's formula (check an introductory calculus text for Taylor's formula):

$$F_X(x + \Delta x) - F_X(x) \approx f_X(x)\Delta x \qquad \text{(for } \Delta x \text{ small)}$$

so that, from Eq. (1.38) we have for small $\Delta x$

$$f_X(x)\Delta x \approx \Pr(x \leq X \leq x + \Delta x). \tag{1.40}$$

Also, by the fundamental theorem of calculus we have

$$F_X(x) = \int_{-\infty}^{x} f_X(z)\, dz \tag{1.41}$$

and more generally

$$F_X(b) - F_X(a) = \int_a^b f_X(x)\,dx. \tag{1.42}$$

Given any function $g(x)$ we define the expectation of $g(X)$ by

$$E\{g(X)\} = \int_{-\infty}^{\infty} g(x)f(x)\,dx. \tag{1.43}$$

In particular, the mean and variance of $X$ are given by

$$\mu_X = E\{X\} = \int_{-\infty}^{\infty} xf(x)\,dx \tag{1.44}$$

$$\sigma_X^2 = E\{(X-\mu_X)^2\} = E\{X^2\} - \mu_X^2. \tag{1.45}$$

Equations (1.43)–(1.45) are analogous to Eqs. (1.15)–(1.17) for the discrete case. Note that the summation symbol $\Sigma_k$ in the discrete case is replaced by an integral symbol $\int_{-\infty}^{\infty}$ (recall from calculus that integration is a kind of "continuous summation") and that the discrete probabilities $p_k$ are replaced by the continuous analog $f(x)\,dx$. There are many pairs of analogous formulas for discrete and continuous random variables. For example, by letting $x \to +\infty$ as in Eq. (1.41), we obtain

$$\int_{-\infty}^{\infty} f_X(x)\,dx = F_X(+\infty) = 1 \tag{1.46}$$

which says that the probability that $X$ takes on some value is equal to one.

We now show how continuous random variables arise in a natural way by considering the time until the first event of a Poisson process. Given a Poisson process with rate parameter $\lambda$, let $T$ denote the time from 0 until the first event occurs. Then $T$ is a continuous random variable whose cumulative distribution function $F_T(t)$ is found as follows:

$$\begin{aligned} F_T(t) &= \Pr(T \le t) \\ &= 1 - \Pr(T > t) \\ &= 1 - \Pr(\text{ no event in } [0,t]) \\ &= 1 - e^{-\lambda t}, \end{aligned}$$

i.e.

$$F_T(t) = 1 - e^{-\lambda t} \qquad (t \geq 0). \tag{1.47}$$

This distribution is called the exponential distribution. The corresponding density $f_T(t)$ is obtained by differentiation

$$f_T(t) = \frac{d}{dt} F_T(t) = \lambda e^{-\lambda t}. \tag{1.48}$$

The mean of $T$, i.e. the average waiting time until the first event, is

$$E\{T\} = \int_0^\infty t f_T(t)\, dt = \frac{1}{\lambda}. \tag{1.49}$$

This is also intuitively reasonable: if $\lambda$ is the average rate of occurrence of events, then the average time between events is $1/\lambda$.*

For the variance of the exponential distribution, we need first to calculate

$$E\{T^2\} = \int_0^\infty t^2 \lambda e^{-\lambda t}\, dt.$$

Let's do this by Feynman's method, first noting that

$$te^{-\lambda t} = -\frac{\partial}{\partial \lambda} e^{-\lambda t}, \quad \text{and} \quad t^2 e^{-\lambda t} = \frac{\partial^2}{\partial \lambda^2} e^{-\lambda t}.$$

Hence

$$\begin{aligned} \int_0^\infty t^2 \lambda e^{-\lambda t} &= \lambda \frac{\partial^2}{\partial \lambda^2} \int_0^\infty e^{-\lambda t}\, dt \\ &= \lambda \frac{\partial^2}{\partial \lambda^2} \left(\frac{1}{\lambda}\right) = \frac{2}{\lambda^2}. \end{aligned}$$

Therefore, we obtain

$$\sigma_T^2 = E\{T^2\} - \mu_T^2 = \frac{1}{\lambda^2}. \tag{1.50}$$

The coefficient of variation of the exponential distribution is exactly 1. (How is this interpreted?)

We have defined $T$ as the waiting time from 0 to the first event. Note, however, that since the time origin is completely arbitrary,

* Eq. (1.49) can be proved by integration by parts, but a better method involves a differentiation with respect to $\lambda$ (see the next paragraph).

$T$ can be considered as the time from "now" until the next event. In particular, $T$ can be considered as the time between successive events. Hence $\mu_T = 1/\lambda$ is the average time between events.

An important assumption underlying the Poisson process, and hence the Poisson distribution, is that the occurrence (or non-occurrence) of an event at some time $t = t_0$ has no influence on the future rate of occurrence of events. Thus

$$\begin{aligned}
\Pr(T > t + t_0 \mid T > t_0) &= \frac{\Pr(T > t + t_0)}{\Pr(T > t_0)} \text{(explain!)} \\
&= \frac{e^{-\lambda(t+t_0)}}{e^{-\lambda t_0}} \\
&= e^{-\lambda t} \\
&= \Pr(T > t)
\end{aligned}$$

which says that the distribution of the waiting time from $t = t_0$ is the same as from $t = 0$. This is sometimes called the memoryless property of the exponential distribution.

In applications to search, this assumption means intuitively that the objects of search are neither "clumped" (so that finding one object increases the chance of finding another), nor "dispersed." In this sense, the Poisson model is said to describe search for "purely randomly distributed" objects. Also, search does not reduce $\lambda$, e.g. by depleting the objects of search.

In nature it is unusual to find plants or animals randomly dispersed, and the Poisson distribution seldom provides a good model. In such cases, an alternative distribution, such as the negative binomial, may be preferred. These are discussed in the next appendix.

## Appendix 1.3
## Some Other Probability Distributions

In this appendix we describe some other probability distributions that are commonly encountered in biology, and which will be useful later in this book. A much more complete list is given in Karlin and Taylor (1977).

### THE BINOMIAL DISTRIBUTION

Consider an experiment with only two outcomes, "success" $S$, which occurs with probability $p$, and "failure" $F$, which occurs with probability $q = 1 - p$:

$$\Pr(S) = p, \qquad \Pr(F) = q = 1 - p.$$

Now consider a sequence of $k$ repetitions of the same experiment; the outcome of a typical such sequence will be denoted by the $k$-tuple $(S, F, S, S, \ldots)$. The probability that a given response will occur is

$$\Pr(S, F, S, S, \ldots) = p^j q^{k-j}$$

where $j$ denotes the number of $S$'s. For example, with $k = 5$

$$\Pr(S, S, F, S, F) = p^3 q^2.$$

Often we may not be interested in the *order* in which successes and failures occur, but only in the number of successes. Let $X_k$ denote the number of successes in a sequence of $k$ repetitions of the experiment; thus $X_k$ is a discrete random variable which can assume any of the values $j = 0, 1, \ldots, k$. We wish to calculate the probability distribution for $X_k$, that is

$$\Pr(X_k = j) \qquad j = 0, 1, \ldots, k.$$

Clearly

$$\Pr(X_k = k) = p^k \quad \text{and} \quad \Pr(X_k = 0) = q^k.$$

The remaining possibilities can be illustrated by the following case:

$$\Pr(X_k = 1) = \Pr(S, F, \ldots, F) + \Pr(F, S, F, \ldots, F) + \cdots + \Pr(F, F, \ldots, S)$$

where each sequence on the right contains exactly one $S$. Now all the terms on the right have the same value, namely $pq^{k-1}$, and there are exactly $k$ of them. Hence

$$\Pr(X_k = 1) = kpq^{k-1}.$$

Similarly, the probability of $j$ successes in $k$ repetitions of the experiment will be the sum of terms all having the same value

$p^j q^{k-j}$. The question is, how many terms are there, each of the form $\Pr(S, S, F, S, \ldots)$, with exactly $j$ $S$'s? The answer is given by the "number of combinations of $k$ objects, taken $j$ at a time," which is given by

$$\binom{k}{j} = \frac{k!}{j!(k-j)!}.$$

The expression $\binom{k}{j}$ is called the *binomial coefficient* (sometimes denoted by ${}_kC_j$), because of its prominence in the binomial theorem:

$$(a+b)^k = \sum_{j=0}^{k} \binom{k}{j} a^j b^{k-j}.$$

For example, the reader may wish to check that $\binom{5}{2} = 10$, which is exactly the total number of sequences of length 5 containing exactly 2 $S$'s.

The final result is that

$$\Pr(X_k = j) = \binom{k}{j} p^j q^{k-j} \qquad j = 0, 1, \ldots, k. \tag{1.51}$$

This is called the *binomial distribution*. As a check, we have

$$\begin{aligned} \sum_{j=0}^{k} \Pr(X_k = j) &= \sum_{j=0}^{k} \binom{k}{j} p^j q^{k-j} \\ &= (p+q)^k \quad \text{by the binomial theorem} \\ &= 1 \quad \text{since } p+q=1. \end{aligned}$$

We suggest that readers develop for themselves an algorithm for calculating $\Pr(X_k = j)$ in (1.51) successively for $j = 0, 1, 2, \ldots, k$.

## Mean and Variance

In order to calculate $\mu_X$ we first ignore the fact that $q = 1 - p$, and calculate

$$\sum_{j=0}^{k} j\binom{k}{j}p^j q^{k-j} = p\frac{\partial}{\partial p}\sum_{j=0}^{k}\binom{k}{j}p^j q^{k-j}$$
$$= p\frac{\partial}{\partial p}(p+q)^k$$
$$= kp(p+q)^{k-1}.$$

Now we can put $p+q=1$ to obtain

$$\mu_X = kp. \tag{1.52}$$

Similarly $E\{X^2\} = kp + k(k-1)p^2$, so that

$$\sigma_X^2 = kpq. \tag{1.53}$$

The variance of the binomial distribution is less than the mean.

### THE NEGATIVE BINOMIAL DISTRIBUTION

An important distribution in ecology is the *negative binomial distribution*, defined by

$$p_j = \frac{\Gamma(j+k)}{j!\Gamma(k)}\left(\frac{k}{k+m}\right)^k\left(\frac{m}{k+m}\right)^j \qquad j = 0,1,2,\ldots \tag{1.54}$$

which depends on two positive parameters, $k$ and $m$. Here the *gamma* function $\Gamma(x)$ is defined by

$$\Gamma(x) = \int_0^\infty t^{x-1}e^{-t}\,dt \qquad (x > 0). \tag{1.55}$$

If you are not familiar with it, the gamma function is an extension of the factorial; when $n$ is an integer

$$\Gamma(n) = (n-1)! \qquad n = 1,2,3,\ldots \tag{1.56}$$

In order to see this, note first that

$$\Gamma(1) = \int_0^\infty e^{-t}\,dt = 1.$$

Next, integration by parts shows that

$$\Gamma(x+1) = x\Gamma(x). \tag{1.57}$$

Hence

$$\Gamma(2) = \Gamma(1) = 1$$
$$\Gamma(3) = 2\Gamma(2) = 2!$$

and so on.

The binomial expansion of $(1-\xi)^{-k}$ can be used to show that $\sum_0^\infty p_j = 1$ (hence the name, negative binomial). The mean and variance of the negative binomial distribution are:

$$\mu = m, \qquad \sigma^2 = m + \frac{m^2}{k}. \tag{1.58}$$

(We will not test the reader's patience by deriving these.)

Note that, in contrast to the binomial distribution, the variance of the negative binomial distribution is larger than the mean. The parameter $k$ is often referred to as the "contagion" or "overdispersion" parameter, with small values of $k$ corresponding to highly "contagious" or "overdispersed" distributions (see Pielou 1977 for applications to the spatial distribution of organisms), since if $k \ll 1$, $\sigma^2 \gg m$.

A computational algorithm for the negative binomial distribution is given by

$$p_0 = \left(\frac{k}{k+m}\right)^k$$
$$p_j = \frac{j+k-1}{j} \cdot \frac{m}{k+m} \cdot p_{j-1} \qquad j = 1, 2, \ldots$$

This follows easily from (1.54).

## Some Continuous Distributions

The simplest continuous probability distribution is the *uniform distribution* on an interval $a \le x \le b$, with density $f_X$ given by (see Figure 1.5)

$$f_X(x) = \begin{cases} \frac{1}{b-a} & \text{if } a < X < b \\ 0 & \text{elsewhere.} \end{cases} \tag{1.59}$$

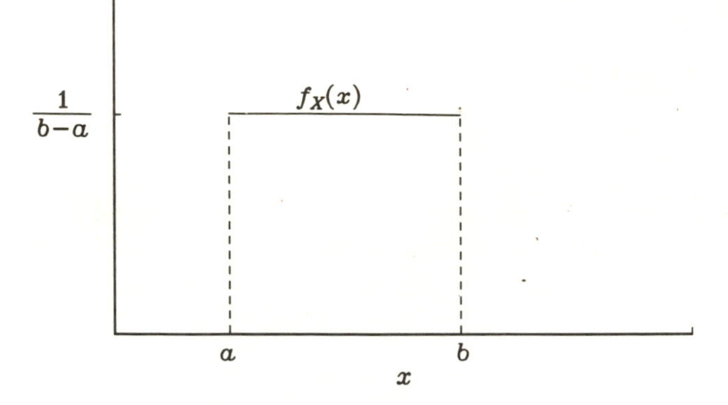


Figure 1.5 The density $f_X(x)$ of the uniform distribution.

Thus

$$\int_{-\infty}^{\infty} f_X(x)\,dx = \int_a^b \frac{dx}{b-a} = 1.$$

For example, suppose a needle is dropped on the floor, and let $X$ denote the direction in which it points. It is reasonable to suppose that $X$ has a uniform distribution on the interval $[0, 2\pi]$. In general, the uniform distribution describes a random variable whose values are equally likely for any $x$ in $[a, b]$, but no other values of $x$ are possible.

The mean of the uniform distribution is

$$\int_a^b x f_X(x)\,dx = \frac{1}{b-a}\int_a^b x\,dx = \frac{1}{b-a}\cdot\frac{b^2-a^2}{2} = \frac{1}{2}(a+b).$$

Perhaps the best known of all probability distributions is the *normal* (or Gaussian) *distribution* $N(\mu, \sigma^2)$ defined by its density (Figure 1.6)

$$f_X(x) = \frac{1}{\sqrt{2\pi}\sigma}\exp\Big(-\frac{(x-\mu)^2}{2\sigma^2}\Big). \tag{1.60}$$

As suggested by the notation, the mean and variance of the normal distribution are $\mu$ and $\sigma^2$.

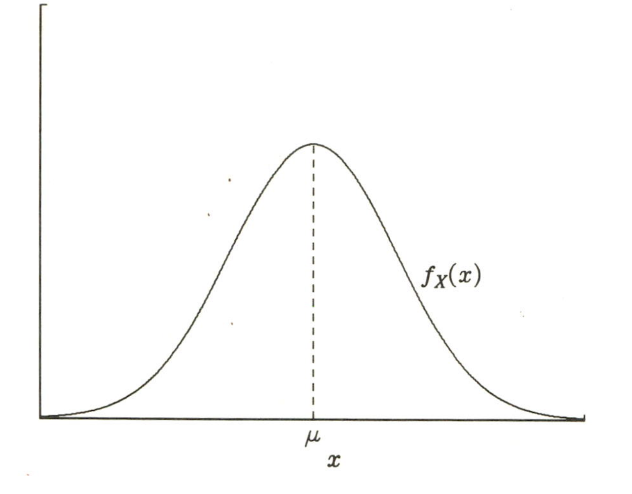


Figure 1.6 The density $f_X(x)$ of the normal distribution.

Many data sets turn out to fit a normal distribution quite well, but note that negative values of $X$ are always possible with a normal distribution. Two popular continuous distributions that apply to non-negative random variables are the lognormal and the gamma distributions.

The random variable $X$ is said to have a *lognormal distribution* if $Y = \ln X$ is normally distributed. In this case one usually works with the log-transformed data, and does not worry about the distribution of $X$ itself.

The density function for the *gamma distribution* has the form

$$f_X(x) = \gamma(x; \nu, \alpha) = \frac{\alpha^\nu}{\Gamma(\nu)} x^{\nu-1} e^{-\alpha x}, \quad x \geq 0 \tag{1.61}$$

(see Figure 1.7).

Note that

$$\begin{aligned} \int_0^\infty f_X(x)\, dx &= \frac{\alpha^\nu}{\Gamma(\nu)} \int_0^\infty x^{\nu-1} e^{-\alpha x}\, dx \\ &= \frac{\alpha^\nu}{\Gamma(\nu)} \int_0^\infty \left(\frac{t}{\alpha}\right)^{\nu-1} e^{-t} \frac{dt}{\alpha} \quad \text{(putting } \alpha x = t\text{)} \\ &= 1 \qquad \text{by (1.55).} \end{aligned}$$

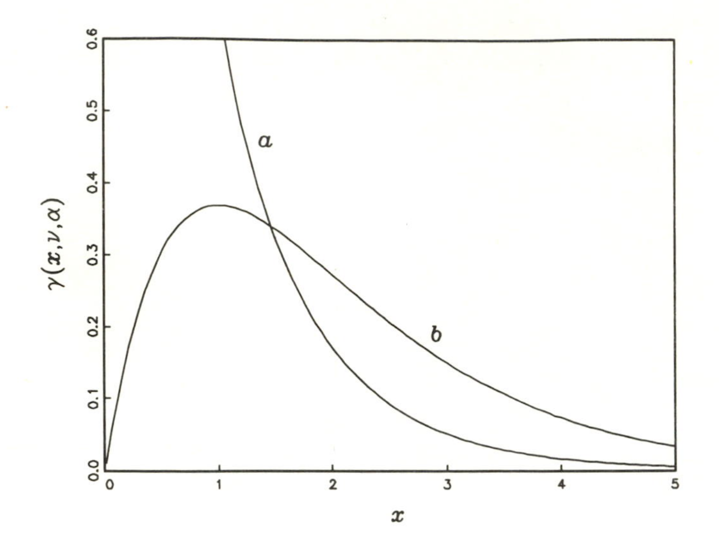


Figure 1.7 The density of the gamma distribution $\gamma(x;\nu,1)$: (a) $\nu = 0.5$, (b) $\nu = 2$.

The mean and variance of the gamma distribution are

$$\mu = \frac{\nu}{\alpha}, \qquad \sigma^2 = \frac{\nu}{\alpha^2}. \tag{1.62}$$

These results can be obtained using another simple computation

$$\begin{aligned}
\mu &= \int_0^\infty f_X(x)\,dx = \frac{\alpha^\nu}{\Gamma(\nu)} \int_0^\infty x^\nu e^{-\alpha x}\,dx \\
&= \frac{\alpha^\nu}{\Gamma(\nu)} \frac{\Gamma(\nu+1)}{\alpha^{\nu+1}} \int_0^\infty \frac{\alpha^{\nu+1}}{\Gamma(\nu+1)} x^{\nu+1-1} e^{-\alpha x}\,dx \\
&= \frac{\nu}{\alpha}
\end{aligned}$$

because by (1.57) $\Gamma(\nu+1) = \nu\Gamma(\nu)$, and because the last integral equals 1, since the integrand is a probability distribution, namely

$\gamma(x; \nu + 1, \alpha)$, and thus its integral must be 1. This explains why we did the calculation in such an apparently strange way. (As an exercise, show that $\int_0^\infty x^a e^{-bx}\,dx = b^{-1-a}\Gamma(a+1)$.)

The distributions described in this section are summarized in Table 1.1. (Notations have been changed slightly from the text above, in some cases.)

## Appendix 1.4
## Renewal Processes

Consider a process in which events occur at random times $t_1$, $t_2, \ldots$. Let $X_k$ denote the time between the $k^{\text{th}}$ and $k+1^{\text{st}}$ event. If the random variables $X_k$ are independent and identically distributed (i.i.d.), then the process is called a *renewal process*. Such a process "renews" itself (in a probabilistic sense) after each occurrence of an event.

The cumulative distribution of any of the $X_k$ is denoted by

$$F(x) = \Pr(X_k \leq x). \tag{1.63}$$

Consider the number of events $N(t)$ in time $t$, and define

$$M(t) = E\{N(t)\}, \tag{1.64}$$

i.e. $M(t)$ is the average number of events in time $t$. The function $M(t)$ is called the *renewal function*.

The simplest renewal process is the Poisson process itself, for which $F(x) = 1 - e^{-\lambda x}$ and $M(t) = \lambda t$. A useful modification is the Poisson-delay process, in which each event is followed by a fixed delay, or "handling time," $h$. This would be more realistic than the Poisson model, both for the telephone call example and the foraging example. We now have

$$F(x) = \begin{cases} 0 & \text{for } 0 \leq x \leq h \\ 1 - e^{-\lambda(x-h)} & \text{for } x > h. \end{cases}$$

This can be further generalized by allowing $h$ to be random, as for the case of a predator encountering different sizes of prey.

Let $\mu = E\{X\}$ be the average length of one renewal period. Then we expect that the average number of events in time $t$ should be $M(t) = t/\mu$. This is in fact true in the limit as $t \to \infty$:

$$\lim_{t\to\infty} \frac{M(t)}{t} = \frac{1}{\mu}. \tag{1.65}$$

Table 1.1
Some Common Probability Distributions

(a) Discrete

| Name | Parameter | $p_j$ | $\mu_X$ | $\sigma_X^2$ |
|---|---|---|---|---|
| Poisson | $\lambda$ | $\lambda^j e^{-\lambda}/j!$ $(j{=}0,1,2,\ldots)$ | $\lambda$ | $\lambda$ |
| binomial | $N, p$ | $\binom{N}{j} p^j (1-p)^{N-j}$ $(j{=}0,1,\ldots,N)$ | $Np$ | $Npq$ |
| negative binomial | $m, k$ | $\dfrac{\Gamma(k+j)}{j!\Gamma(k)} \left(\dfrac{k}{k+m}\right)^k \left(\dfrac{m}{k+m}\right)^j$ $(j{=}0,1,2,\ldots)$ | $m$ | $m + \frac{m^2}{k}$ |

(b) Continuous

| Name | Parameter | $f_X(x)$ | | $\mu_X$ | $\sigma_X^2$ |
|---|---|---|---|---|---|
| uniform | $a, b$ | $\dfrac{1}{b-a}$ | $(a<x<b)$ | $\dfrac{a+b}{2}$ | $\dfrac{a^2-2ab+b^2}{12}$ |
| exponential | $\lambda$ | $\lambda e^{-\lambda x}$ | $(x>0)$ | $1/\lambda$ | $1/\lambda^2$ |
| normal | $\mu, \sigma$ | $\dfrac{1}{\sqrt{2\pi}\sigma} \exp\left(-\dfrac{(x-\mu)^2}{2\sigma^2}\right)$ | $(-\infty<x<\infty)$ | $\mu$ | $\sigma^2$ |
| gamma | $\alpha, \nu$ | $\dfrac{\alpha^\nu}{\Gamma(\nu)} x^{\nu-1} e^{-\alpha x}$ | $(x>0)$ | $\nu/\alpha$ | $\nu/\alpha^2$ |

This equation, called the *renewal theorem*, says that the long-term average rate of event occurrence equals $\frac{1}{\mu}$. The renewal theorem is mathematically nontrivial, and has been rediscovered by many authors; see Karlin and Taylor (1977) for a fuller discussion.

For the case of the Poisson-delay process with fixed handling time $h$ we have

$$E\{X\} = h + \frac{1}{\lambda}$$

so that

$$\frac{M(t)}{t} \approx \frac{1}{h + 1/\lambda} \qquad \text{for large } t\,. \tag{1.66}$$

If the delay $h$ is also random, with mean $\mu_h$, then we have

$$E\{X\} = \mu_h + \frac{1}{\lambda}$$

and (1.66) becomes

$$\frac{M(t)}{t} \approx \frac{1}{\mu_h + 1/\lambda} \qquad \text{for large } t. \tag{1.67}$$

Equation (1.67) and its extensions are the foundations of classical foraging theory, including Holling's disc equation (Holling 1965), the "marginal value theorem" (Charnov 1976), and diet selection theory (Schoener 1971).

As an easy extension of this analysis, consider the following foraging model:

(a) search is a Poisson process with rate parameter $\lambda$;
(b) handling time is a random variable $h$ with mean $\mu_h$;
(c) food size is a random variable $B$ with mean $\mu_B$.

Let $Q(t)$ denote total food consumed in time $t$; the long-term average rate of food consumption can then be shown to equal

$$\lim_{t\to\infty} \frac{E\{Q(t)\}}{t} = \frac{\mu_B}{\mu_h + 1/\lambda}. \tag{1.68}$$

Note that the right side of Eq. (1.68) is equal to

$$\frac{E\{\text{food consumed per encounter}\}}{E\{\text{time to find and consume food}\}}. \tag{1.69}$$

It was recently suggested (Templeton and Lawlor 1981) that this quotient was fallacious and should be replaced by

$$E\left\{\frac{\text{food consumed per encounter}}{\text{time to find and consume food}}\right\}. \tag{1.70}$$

As pointed out by Turelli et al. (1982), however, this "fallacy" is itself a fallacy. Eq. (1.70) would represent the average rate of food intake for a large number of foragers, each making a single encounter. This is not the same as the long-term rate for a single forager, which is correctly represented by (1.69).

Finally, consider a model of foraging with patch *depletion*. Let $Z_t$ denote the number of prey remaining in the patch:

$$Z_t = Z_0 - N_t.$$

It might be reasonable to suppose that the Poisson search parameter $\lambda$ is proportional to $Z_t$, i.e. $\lambda$ depends on time $t$. Notice that this is *not* a renewal process, since the distribution of detection times is not the same for each detection. Search models with depletion are usually much more difficult to analyze than search models which ignore depletion (Mangel and Clark 1983).

# 2 Patch Selection

We begin the discussion of dynamic modeling by considering the problem of "patch selection" in a relatively abstract context. This is done so that the main concepts can be considered and illuminated without undue concern for biological details—in later chapters we will treat biological details with considerable care. Additional information regarding the formulation of dynamic optimization models of behavior is given in Chapter 8.

In the main part of this chapter (Sections. 2.3–2.5) we treat the simplest possible patch selection model. However, lest the reader get the false impression that the basic modeling framework is extremely limited, we also discuss some important variations of the simplest model. Several further elaborations are discussed in the appendices to this chapter. The latter material can be skipped at a first reading, but we suggest that the reader at least glance through the appendices to get a feeling for how flexible the dynamic approach really is. Familiarity with the methods described in the appendices is essential for anyone hoping to develop his or her own dynamic behavioral models.

## 2.1 Patch Selection as a Paradigm

If "foraging theory," understood in the usual modern context, has a clear starting point, it can probably be traced to two papers which appeared together in the *American Naturalist* in 1966 (MacArthur and Pianka 1966, and Emlen 1966) on the optimal use of a patchy environment. In these papers the concern is with behavior that in some sense optimizes a rate of energy return to the foraging individual.

We will think of "patch selection" in a much broader context—although certainly the search for food is a crucial aspect of animal behavior—in which "patches" correspond to activity choices. These activity choices are characterized by rewards to the individual (e.g. food obtained or eggs laid), costs to the individual (e.g. time spent, metabolic costs), and risks to the individual (e.g. the possibility of predation). The problem of "optimal patch use" thus reduces to the following question: What is the best way to trade off the rewards with the costs and the risks? Before showing how to address this question using dynamic state variable models, we will discuss a number of examples from the recent literature, which resemble the patch selection problem as understood in this sense.

## 2.2 Biological Examples

### Birds

A sampling of the many studies concerning the response of birds to patches of resources includes the following. Bryant and Turner (1982) studied the central place foraging of house martins (*Delicon urbica*), sand martins (*Riparia riparia*), and swallows (*Hirundo rustica*). The birds were provided with food sources at various distances from the nest, and the response of the birds—measured in terms of amount of prey carried back to the nest as a function of distance from the nest to the patch of resources—was studied. In this case, the tradeoff is between a reward (food) and a set of costs (travel time and energy). Similar kinds of experiments were performed by Carlson and Moreno (1981) using breeding wheatears (*Oenanthe oenanthe*) and by Hegner (1982) using white-fronted bee-eaters (*Merops bullockoides*).

Caldwell (1986) studied the foraging behavior of little blue herons (*Egretta caerulea*) in the tropics. These birds could exploit one of two "patches"—either mangrove forests or reefs. The mangroves were preferable to the reefs in a number of ways: they were closer to nest sites, the energy cost of finding food was less since there were resting spots, and the prey items in the mangroves were somewhat larger. The risk of foraging in the mangroves, however, was considerably greater in some years, due to the presence of hawks (*Buteogallus anthracinus*). Caldwell discovered that in years of heavy attacks by predators, the herons shifted to foraging in the poorer habitat "under unfavorable climatic conditions, and under thermoregulatory stress." One of the major points of Caldwell's work was the study of group formation and flocking (see Chapter 3 of this book). Another aspect, not so clearly formulated by Caldwell, was the "informational question." That is, at the start of a given year the herons did not know what the hawk predation level would be. It is only by learning about predation risks that the decision to switch to a poorer habitat could be made (see Chapter 9 for a discussion of informational problems in dynamic modeling of behavior).

Caraco et al. (1980) provide an empirical demonstration of risk sensitive foraging preferences in yellow-eyed juncos (*Junco phaeonotus*). They showed that the birds responded not only to the mean of food rewards in the patches, but also to the variance. We will show later in this chapter how dynamic modeling can shed light on this kind of behavior.

Grubb and Greenwald (1982) studied the response of sparrows (*Passer domesticus*) to different combinations of predation risk and energy cost. In particular, they studied the shifting between colder, safer feeding sites and warmer, riskier feeding sites.

Lima (1984, 1985a, 1985b) studied the behavior of a number of different kinds of birds in a variety of different "patch type" settings. In Lima (1984) downy woodpeckers (*Picoides pubsecens*) were presented with artificial patches consisting of logs with holes drilled in them, some portion of the holes being filled with seeds. In these experiments, Lima was concerned again with informational issues in patch selection, particularly the question of how many holes in a patch should be sampled before deciding whether the patch was "good" or "bad" (determined by the number of seeds in the patch). In another experiment, using starlings (*Sturnus vulgaris*), patches either contained no prey or a fixed number of prey, and the starling had to determine the level of effort to put into a patch before deciding that it was empty (Lima 1985a). In Lima (1985b), the "predation–starvation risk" question was studied using black-capped chickadees (*Parus atricapillus*), which were forced to make a decision concerning foraging efficiency (measured in energy intake rate) and time exposed to a predator.

### Fish

Many experiments on fish behavior are also concerned with patch selection. Fraser and Cerri (1982) and Cerri and Fraser (1983) conducted experiments on predation risk and foraging behavior in minnows (*Semotilus atromaculatus* and *Rhinichthys atratulus*) in headwater streams. These streams contained pools and riffles and thus a patchy habitat for the foraging minnows. Predators were adult *S. atromaculatus*, and the foragers were juveniles of both species. Four kinds of patches were constructed: high food, low risk; high food, high risk; low food, low risk; and low food, high risk. The relative distribution of foragers in these patches was studied. Milinski (1986) provides an analysis of this set of experiments and of his own work on pike (*Esox lucius*) and three-spined sticklebacks (*Gasterosteus aculeatus*), in which the tradeoff considered is again the value of food and the risk of predation. Milinski and Heller (1978) developed state variable models, using deterministic optimal control theory, for experiments of this kind.

Pitcher and Magurran (1983) and Pitcher (1986) describe experiments using goldfish (*Carassius auratus*) and minnows, on shoal size, patch profitability (in terms of energy return), and information exchange. Another set of experiments on the value-of-food, predation-risk tradeoff was conducted by E. Werner and his colleagues (Werner and Mittelbach 1981, Werner et al. 1983a,b) using bluegill sunfish (*Lepomis macrochirus*) with daphnia (*Daphnia magna*) as a prey item and large mouth bass (*Micropterus salmoides*) as a predator. They found, for example, that when bass were present, larger sunfish chose to maximize the rate of energy return but smaller sunfish chose to forage in a patch containing vegetation, in which the energy return rate was much lower but the chance of predation was also much lower. Many of these experiments, and others, are summarized in Werner and Gilliam (1984). In Chapter 5 we discuss models of the movements of fish and other aquatic organisms in response to food availability and risks of predation.

### Insects

Large portions of the life of an insect are tied to patch selection problems—looking either for food sites or for oviposition sites. Some typical examples are found in the papers by Courtney (1986), Jones (1977), and Root and Kareiva (1984) concerning the cabbage white butterfly (*Pieris rapae*), Marris et al. (1986) concerning the parasitic wasp *Nemeritis canescens*, Morrison and Lewis (1981) concerning another parasite *Trichogramma pretiosum*, and Whitham (1980) on Pemphigus aphids. These papers, as well as the ones which will be discussed in detail in Chapter 4, show the need to analyze the patch selection problem in order to understand insect behavioral decisions.

These examples indicate the widespread interest and variety of "patch selection" problems and also suggest the need for a unified and simple modeling paradigm. Such an approach is developed in this chapter. In subsequent chapters we apply this paradigm, suitably modified, to a variety of biological situations, paying careful attention to biological detail.

## 2.3 The Simplest State Variable Model

In this section we discuss the patch selection problem using the simplest possible state variable model. We use a single state variable $X(t)$ to characterize the state of the forager at time $t$. At this point, the exact nature and units of $X(t)$ do not need to be specified, but it may be helpful to think of $X(t)$ as the forager's energy reserves at time $t$. We consider a long time interval of length $T$, and assume that the state of the forager changes with each unit increment in time. That is, we wish to relate the value of the state variable $X(t+1)$ at time $t+1$ to its value $X(t)$ at time $t$. In order to relate these two quantities, we need to characterize the patches. We suppose that the patch parameters are:

$\alpha_i$ = Cost per period (measured in terms of decrement of the state variable) of choosing patch $i$;

$\beta_i$ = Probability of predation during any one period if patch $i$ is chosen;

$\lambda_i$ = Probability of finding food during any one period if patch $i$ is chosen; and

$Y_i$ = Increment in the state variable (if food is discovered) in the $i$th patch.

Before going on, it is worthwhile to point out a number of aspects of this model. (Figure 2.1 shows a schematic illustration of the patch selection problem, in which costs and rewards are modeled as food.) First, we are using what is called in the mathematical jargon a *discrete-time model*; that is, we only allow time to be incremented by discrete units. Although time is actually a continuous variable, we stress that it is the use of discrete-time models which makes the dynamic modeling approach discussed in this book so easy to use. Note too that the units of time are not specified, so that by choosing the units properly we can model a wide variety of circumstances. For example, when studying the hunting behavior of lions (Chapter 3), the appropriate unit of

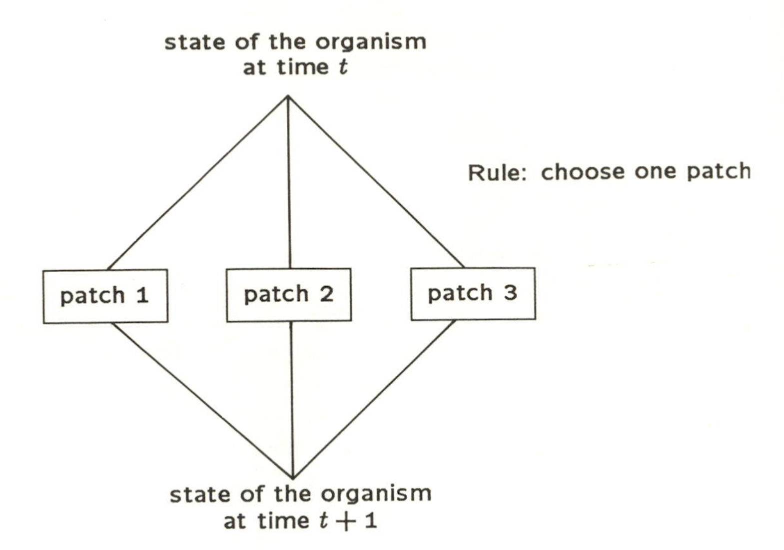


Figure 2.1 Schematic illustration of the patch selection problem. At time $t$, the organism must choose between one of three patches. Each patch is characterized by a set of four parameters: energetic cost of foraging in the patch ($\alpha$), probability of predation in the patch ($\beta$), probability of finding food in the patch ($\lambda$), and energetic value of food in the patch ($Y$). If the organism survives predation, it begins period $t+1$ with a state determined both by which patch was chosen and by what happened in the patch.

time might be one day, while for small birds foraging in the winter an appropriate unit of time might be five minutes. Second, note that all of the "patch parameters," $\alpha$, $\beta$, $\lambda$, and $Y$, are assumed to be constant in time, but varying between patches. This assumption is easily relaxed if one knows that the patch parameters are time dependent. The modeling approach that we develop is extremely flexible and can readily accommodate such changes. Consequently, the modeling framework is easily adapted to actual field or laboratory studies, as we hope to demonstrate in Chapters 3–7.

To return to the dynamic modeling, we will first consider a non-breeding period, so that the "objective" of the animal is simply survival. That is, we identify fitness with the probability of survival. In particular, we define a *lifetime fitness function* $F(x,t,T)$ in the following manner:

$F(x,t,T)$ = maximum value of the probability that the forager survives from period $t$ to period $T$, given that at period $t$ the forager is alive, and the value of the state variable $X(t)$ is $x$. (2.1)

We can write the right-hand side of Eq. (2.1) much more conveniently by using the notation introduced in Chapter 1:

$$F(x,t,T) = \max \Pr(\text{survive from } t \text{ until } T \mid X(t) = x). \quad (2.1')$$

In addition, we assume that there is a critical value of the state variable, denoted by $x_c$, such that if $X(t)$ falls below this critical value at any time, then the forager dies:

$$F(x,t,T) = 0 \text{ for } x \leq x_c. \quad (2.2)$$

What do we mean by "maximum probability" in Eq. (2.1)? Recall that we are trying to predict optimal behavior, i.e. in this model, the optimal patch choice. By definition the optimal patch choices are those that lead to the maximum probability of survival. We define $F(x,t,T)$ to be equal to this maximum probability of survival, assuming that the optimal patch is chosen in each period $t, t+1, \ldots, T$. Of course we have not yet shown how to calculate $F(x,t,T)$, but we will do so shortly. At the same time that $F(x,t,T)$ is calculated, the optimal decisions are also determined.

At this point, it is extremely important to define clearly how we measure time and assess changes in the state variable. The reader may find Figure 2.2 helpful. We will assume that the state of the forager is assessed *at the start of each period t*. This means that if $X(t) > x_c$, then the forager is alive, and otherwise it is dead. Next, we assume that for $t < T$, changes in the state variable occur *between t and t + 1* and that these changes are determined by chance events associated with predation and with finding food. The probabilities associated with these chance events are determined by the patch chosen for the interval between $t$ and $t+1$; in each such period a particular patch is used, and the forager can change patches from one period to the next.

For the terminal period $T$ we are only concerned with whether the forager is alive or dead. Thus one value of $F(x,t,T)$ is already known:

$$F(x,T,T) = \begin{cases} 1 & \text{if } x > x_c \\ 0 & \text{if } x \leq x_c. \end{cases} \quad (2.3)$$

We are now ready to derive the so called *dynamic programming equation* that $F(x,t,T)$ satisfies. To do this, assume that at some

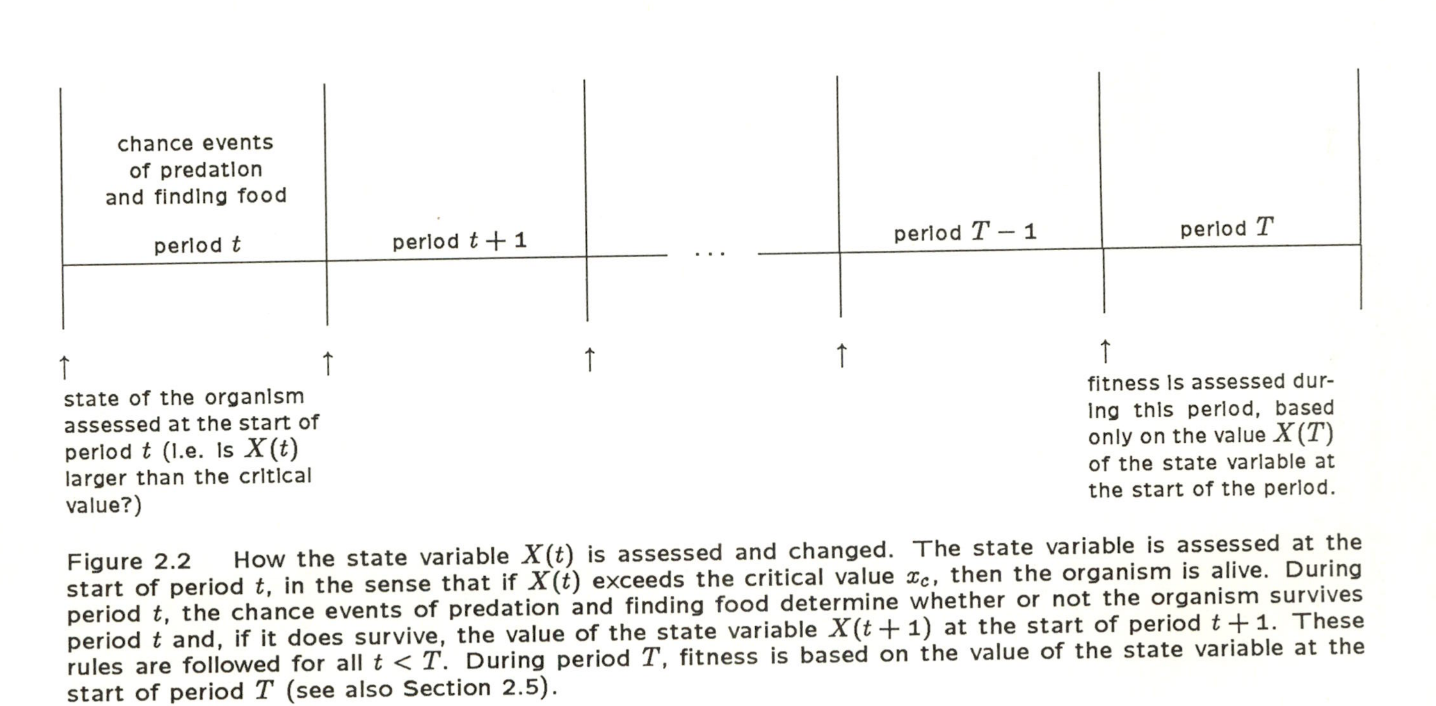


Figure 2.2 How the state variable $X(t)$ is assessed and changed. The state variable is assessed at the start of period $t$, in the sense that if $X(t)$ exceeds the critical value $x_c$, then the organism is alive. During period $t$, the chance events of predation and finding food determine whether or not the organism survives period $t$ and, if it does survive, the value of the state variable $X(t+1)$ at the start of period $t+1$. These rules are followed for all $t < T$. During period $T$, fitness is based on the value of the state variable at the start of period $T$ (see also Section 2.5).

intermediate period, $t$, the value of the state variable is $X(t) = x$. Suppose that the forager chooses patch $i$. Then, if the forager survives predation (which occurs with probability $1 - \beta_i$), the new value of the state variable becomes

$$X(t+1) = \begin{cases} x - \alpha_i + Y_i & \text{with probability } \lambda_i \\ x - \alpha_i & \text{with probability } 1 - \lambda_i. \end{cases} \tag{2.4}$$

Two comments are worth making at this point. First, what happens if $x - \alpha_i$ or $x - \alpha_i + Y_i$ are less than the critical value $x_c$? We adopt the assumption that if $X(t+1)$ falls below $x_c$, then $X(t+1)$ is set equal to $x_c$; the forager is then dead. (The reader may find it helpful to assume, unless there is good reason to do otherwise, that the critical level is $x_c = 0$.) Second, in many biological situations the state variables are constrained. For example, suppose that the forager has a capacity constraint $C$, in the sense that $X(t)$ can never exceed $C$. Thus

$$X(t) \leq C \qquad \text{for all } t.$$

We can handle both the capacity constraint and the critical value constraint by introducing a special notation, the "chop" function:

$$\text{chop}(x; a, b) = \begin{cases} b & \text{if } x > b \\ x & \text{if } a \leq x \leq b \\ a & \text{if } x < a. \end{cases} \tag{2.5}$$

In other words, chop$(x; a, b)$ simply chops the value of the variable $x$ at the lower and upper limits $a$ and $b$. With this notation we now rewrite Eq. (2.4) in a more precise form including the constraints:

$$X(t+1) = \begin{cases} \text{chop}(x - \alpha_i + Y_i; x_c, C) & \text{with probability } \lambda_i \\ \text{chop}(x - \alpha_i; x_c, C) & \text{with probability } (1 - \lambda_i). \end{cases} \tag{2.6}$$

The reader should take the time to check that Eq. (2.6) is exactly what we need; for example, if the forager finds food $Y_i$, then it will eat it all unless this results in exceeding the capacity (after subtracting the metabolic cost), and so on.

Having specified the state dynamics, we can now proceed to derive the dynamic programming equation for $F(x, t, T)$. Let us begin with the penultimate period, $t = T - 1$. Assuming that $X(T-1) = x$, suppose that the forager chooses patch $i$ in period

$T-1$. Then (refer to Figure 2.2) the forager will survive period $T-1$ if it is not killed by a predator, i.e. with probability $(1-\beta_i)$. Its state variable will change from $X(T-1)=x$ to $X(T)$ as given by Eq. (2.6). Also, the forager will survive the final period $T$ if and only if $X(T) > x_c$. We can therefore write

$$\begin{aligned} &\Pr(\text{survive from } T-1 \text{ to } T \mid \text{choose patch } i \text{ in period } T-1) \\ &\quad = \Pr(\text{survive period } T-1 \mid \text{choose patch } i \text{ in period } T-1) \\ &\quad\quad \times \Pr(\text{survive period } T \mid \text{choose patch } i \text{ in period } T-1). \end{aligned} \tag{2.7}$$

Now

$$\Pr(\text{survive period } T-1 \mid \text{choose patch } i \text{ in period } T-1) = 1-\beta_i.$$

Furthermore, by the Law of Total Probability, Eq. (1.11)

$$\begin{aligned} &\Pr(\text{survive period } T \mid \text{choose patch } i \text{ in period } T-1) \\ &\quad = \lambda_i \Pr(\text{survive period } T \mid X(T) = x_i') \\ &\quad\quad + (1-\lambda_i)\Pr(\text{survive period } T \mid X(T) = x_i'') \end{aligned}$$

where

$$\left.\begin{aligned} x_i' &= \text{chop}(x-\alpha_i+Y_i; x_c, C) \\ x_i'' &= \text{chop}(x-\alpha_i; x_c, C) \end{aligned}\right\} \tag{2.8}$$

By definition of $F(x,T,T)$ we have

$$\Pr(\text{survive period } T \mid X(T) = x) = F(x,T,T).$$

Substituting these values into (2.7), we conclude that the probability of surviving both periods $T-1$ and $T$ (if patch $i$ is chosen in period $T-1$) equals

$$(1-\beta_i)[\lambda_i F(x_i',T,T) + (1-\lambda_i)F(x_i'',T,T)].$$

By definition $F(x,T-1,T)$ is the maximum of these probabilities, relative to patch choice:

$$F(x,T-1,T) = \max_i (1-\beta_i)[\lambda_i F(x_i',T,T) + (1-\lambda_i)F(x_i'',T,T)]. \tag{2.9}$$

This is the dynamic programming equation for $t = T - 1$; notice that everything on the right side of (2.9) is already known, since $F(x, T, T)$ is given by (2.3). Hence $F(x, T-1, T)$ can now be computed by finding the patch index $i = i^*$ which gives the maximum in Eq. (2.9). Moreover, this also determines the optimal strategy $i^*$ for $t = T - 1$; note that $i^*$ in general will depend on $x$.

It is extremely important that the detailed reasoning that led to Eq. (2.9) be fully understood. The same type of reasoning must be used for every dynamic optimization model in this book, but each new model may have its own peculiarities. We encourage the reader to go through the reasoning again—for example, imagine that you have to explain it to someone else.

Exactly the same line of reasoning applies to any $t < T$, relating fitness in period $t$ to fitness in the next period. We thereby obtain the general form of the dynamic programming equation:

$$F(x,t,T) = \max_i (1 - \beta_i)[\lambda_i F(x_i', t+1, T) + (1 - \lambda_i) F(x_i'', t+1, T)] \tag{2.10}$$

where $x_i'$ and $x_i''$ are given by Eqs. (2.8). Of course, Eq. (2.10) is only used for $x > x_c$; we always have

$$F(x,t,T) = 0 \quad \text{if} \quad x \le x_c. \tag{2.11}$$

Since Eq. (2.10) is the fundamental equation of our patch selection paradigm, let us now try to understand it intuitively. First, $F(x,t,T)$ is the maximum probability of surviving from $t$ to $T$, given that $X(t) = x$. In period $t$ the forager must choose some patch $i$. This will affect both its current probability of survival $1-\beta_i$, and its state variable $X(t+1)$, which can assume one of two values, $x_i'$ or $x_i''$. [Recall that if either of these falls below $x_c$ the forager dies of starvation; this is automatically taken care of by Eq. (2.11).] Finally, the maximum probability of surviving from $t+1$ to $T$, if $X(t+1) = x'$, is by definition $F(x', t+1, T)$. With probability $\lambda_i$ (i.e. food $Y_i$ is found) this equals $F(x_i', t+1, T)$, and with probability $1 - \lambda_i$ this equals $F(x_i'', t+1, T)$. Thus Eq. (2.10) follows.

In the following section we explain how Eq. (2.10) is used to solve our dynamic optimization problem.

## 2.4 An Algorithm for the Dynamic Programming Equation

We will now assume that the reader has access to some kind of computer—a moderate desktop micro is all that is needed for nearly all of the computations discussed in this book and for all of the computations discussed in this chapter. Any reader who has never programmed a computer should refer to the Addendum to Part I before attempting to write a program.

To begin, we assume that in addition to time being measured discretely, the state variable also takes integer values. Since no units are specified, we can measure $x$, $\alpha_i$, and $Y_i$ as multiples of the smallest of these units. The use of integer values for time and the state variable means that the algorithm is extremely simple. In Chapter 8, we show how to extend the algorithm presented in this chapter to noninteger state variables.

Now let us write down the dynamic programming equations (2.10) and (2.11) again:

$$F(x,t,T) = \begin{cases} \max\limits_i (1-\beta_i)[\lambda_i F(x_i',t+1,T) & \\ \quad +(1-\lambda_i)F(x_i'',t+1,T)] & \text{if } x > x_c \\ 0 & \text{if } x \le x_c \end{cases} \tag{2.12}$$

where

$$\left.\begin{aligned} x_i' &= \text{chop}(x-\alpha_i+Y_i;x_c,C) \\ x_i'' &= \text{chop}(x-\alpha_i;x_c,C) \end{aligned}\right\} \tag{2.13}$$

These equations lead to an algorithm for computing the optimal patch selection strategy; the algorithm also gives the values of the lifetime fitness function $F(x,t,T)$.

The algorithm works as follows. First the values of $F(x,T,T)$ are given by Eq. (2.3):

$$F(x,T,T) = \begin{cases} 1 & \text{if } x > x_c \\ 0 & \text{if } x \le x_c. \end{cases}$$

Hence $F(x,T-1,T)$ can be computed by using Eq. (2.12). Having computed $F(x,T-1,T)$ for all $x$, we can then use Eq. (2.12) to compute $F(x,T-2,T)$. Using this process of iteratively solving Eq. (2.12) for $t = T-1, T-2, \ldots$, we finally end with $F(x,1,T)$. Furthermore, each iteration also tells us what the optimal patch selection strategy $i^*$ is; in general it will turn out that $i^*$ will depend both on the current state $x$ and the time $t$.

In order to solve the dynamic programming equation, you will need the following in your computer program:

(1) Two vectors (or arrays) of dimension $C$ (the capacity expressed in terms of basic units of $x$) representing $F(x,t,T)$ and $F(x,t+1,T)$. For purposes of illustrating the algorithm we will call $F0(x)$ the vector corresponding to $F(x,t,T)$ and $F1(x)$ the vector corresponding to $F(x,t+1,T)$.

(2) Vectors that characterize the patch parameters. Let us suppose that there are a total of $M$ possible patches. Then you will need four vectors of dimension $M$. The first one, $A$, will correspond to the values of $\alpha$. That is, the $i$th element in the vector $A$ will be $\alpha_i$. The second vector, $B$, will correspond to the values of $\beta$, with the same interpretation of the $i$th element. The third and fourth vectors, $L$ and $Y$, will correspond to the values of $\lambda$ and $Y$ in the various patches.

(3) A vector $D$ of dimension $C$ that keeps track of the optimal patch index $i^* = i^*(x)$. After each iteration, $D(x)$ will be printed out, together with the vector $F0(x)$.

That is all you need. Here is the algorithm.

*Step 1.* Initialize the vector $F1(x)$, corresponding to the value of $F(x,T,T)$. To do this, cycle through values of $x$, starting at $x = x_C$ and going to $x = C$. Note that this is particularly easy because discrete values of the state variable are used. If $x > x_C$, the critical value, then set $F1(x) = 1$. Also, set $F1(x_C) = 0$. After this is done, set $t = T$.

*Step 2.* This is the heart of the algorithm. Cycle over $x$ from $x = x_C + 1$ to $x = C$ again. This time, however, at each value of $x$ cycle also over patches. To do this, hold $x$ fixed and cycle over $i = 1$ to $M$; compute

$$V_i = (1 - \beta_i)\{\lambda_i F1(x_i') + (1 - \lambda_i) F1(x_i'')\}$$

where $x_i'$ and $x_i''$ are given by Eq. (2.13). Determine the largest of all the $V_i$ and call it $VM$. Set $F0(x) = VM$ and set $D(x)$ equal to the value of $i$ which gives the largest value, $VM$. Continue cycling over all values of the state variable $x$. Also set $F0(x_c) = 0$. This completes one step of the "iteration" procedure, and computes $F0(x) = F(x,t,T)$ in terms of $F1(x) = F(x,t+1,T)$, according to the dynamic programming equation (2.12).

*Step 3.* Now print out the value of $t$ and the values of $D(x)$ and $F0(x)$ for $x = x_c + 1$ to $C$.

*Step 4.* Once again cycle over values of $x$. For each value of $x$, replace the current value of $F1(x)$ by $F0(x)$. This step "updates" the fitness function, putting $F1(x) = F(x,t,T)$.

*Step 5.* Replace $t$ by $t - 1$. If $t > 0$, then go back to Step 2 and repeat the entire process over again. Otherwise stop.

As an example, consider a simple problem with three patches. Suppose that the patch parameters are the following:

| Patch # | $\beta$ | $\alpha$ | $\lambda$ | $Y$ |
|---|---|---|---|---|
| 1 | 0 | 1 | 0 | 0 |
| 2 | .004 | 1 | .4 | 3 |
| 3 | .020 | 1 | .6 | 5 |

With these parameters, patch 1 is a "safe" patch, since there is no chance of predation, but there is also no chance of finding food. Patch 3 is the riskiest patch in terms of predation, but the expected return $\lambda_3 Y_3 = 3.0$ exceeds the expected return $\lambda_2 Y_2 = 1.2$ in the safer patch 2.

We now recommend that before reading any more of this book, the reader goes to a computer, codes up the algorithm that was just described, and studies the output.* The rest of the discussion in this section is predicated on the assumption that you have successfully coded the algorithm. We stress that this material is best learned by doing, rather than by reading about it.

Other parameter values suggested: $x_c = 3$, $C = 10$, and $T = 10$, 20, or 30.

Table 2.1 shows the program output for this model, with $T = 20$. The following features are worth noting:

(1) At each time $t$, fitness (i.e. the probability of survival) is

* Novices at programming mathematical calculations may wish to read the Addendum to Part I.

Table 2.1
Probability of survival $F(x,t,T)$ and optimal patch choice $i^* = i^*(x,t)$ for the simplest patch-selection model with $T = 20$.

| Energy reserves | $F(x,t,T)$ | $i^*$ | $F(x,t,T)$ | $i^*$ | $F(x,t,T)$ | $i^*$ | $F(x,t,T)$ | $i^*$ |
|---|---|---|---|---|---|---|---|---|
| | $t = 19$ | | $t = 18$ | | $t = 17$ | | $t = 16$ | |
| 4 | 0.588 | 3 | 0.588 | 3 | 0.588 | 3 | 0.588 | 3 |
| 5 | 1.000 | 1 | 0.818 | 3 | 0.818 | 3 | 0.818 | 3 |
| 6 | 1.000 | 1 | 1.000 | 1 | 0.909 | 3 | 0.909 | 3 |
| 7 | 1.000 | 1 | 1.000 | 1 | 1.000 | 1 | 0.944 | 3 |
| 8 | 1.000 | 1 | 1.000 | 1 | 1.000 | 1 | 1.000 | 1 |
| 9 | 1.000 | 1 | 1.000 | 1 | 1.000 | 1 | 1.000 | 1 |
| 10 | 1.000 | 1 | 1.000 | 1 | 1.000 | 1 | 1.000 | 1 |
| | $t = 15$ | | $t = 14$ | | $t = 13$ | | $t = 12$ | |
| 4 | 0.588 | 3 | 0.566 | 3 | 0.566 | 3 | 0.566 | 3 |
| 5 | 0.818 | 3 | 0.818 | 3 | 0.794 | 3 | 0.794 | 3 |
| 6 | 0.909 | 3 | 0.909 | 3 | 0.909 | 3 | 0.888 | 3 |
| 7 | 0.944 | 3 | 0.944 | 3 | 0.944 | 3 | 0.933 | 3 |
| 8 | 0.963 | 2 | 0.963 | 2 | 0.963 | 2 | 0.955 | 2 |
| 9 | 1.000 | 1 | 0.974 | 2 | 0.974 | 2 | 0.966 | 2 |
| 10 | 1.000 | 1 | 1.000 | 1 | 0.980 | 2 | 0.974 | 1 |
| | $t = 11$ | | $t = 10$ | | $t = 9$ | | $t = 8$ | |
| 4 | 0.561 | 3 | 0.556 | 3 | 0.550 | 3 | 0.546 | 3 |
| 5 | 0.790 | 3 | 0.784 | 3 | 0.776 | 3 | 0.769 | 3 |
| 6 | 0.884 | 3 | 0.878 | 3 | 0.871 | 3 | 0.863 | 3 |
| 7 | 0.921 | 3 | 0.914 | 3 | 0.908 | 3 | 0.900 | 3 |
| 8 | 9.945 | 2 | 0.935 | 2 | 0.928 | 2 | 0.921 | 2 |
| 9 | 0.959 | 2 | 0.950 | 2 | 0.941 | 2 | 0.933 | 2 |
| 10 | 0.966 | 1 | 0.959 | 1 | 0.950 | 1 | 0.941 | 1 |
| | $t = 7$ | | $t = 6$ | | $t = 5$ | | $t = 4$ | |
| 4 | 0.541 | 3 | 0.537 | 3 | 0.532 | 3 | 0.527 | 3 |
| 5 | 0.763 | 3 | 0.756 | 3 | 0.750 | 3 | 0.743 | 3 |
| 6 | 0.854 | 3 | 0.848 | 3 | 0.840 | 3 | 0.833 | 3 |
| 7 | 0.891 | 3 | 0.884 | 3 | 0.876 | 3 | 0.869 | 3 |
| 8 | 0.912 | 2 | 0.904 | 2 | 0.897 | 2 | 0.889 | 2 |
| 9 | 0.925 | 2 | 0.917 | 2 | 0.909 | 2 | 0.901 | 2 |
| 10 | 0.933 | 1 | 0.925 | 1 | 0.917 | 1 | 0.909 | 1 |
| | $t = 3$ | | $t = 2$ | | $t = 1$ | | | |
| 4 | 0.523 | 3 | 0.518 | 3 | 0.514 | 3 | | |
| 5 | 0.737 | 3 | 0.730 | 3 | 0.724 | 3 | | |
| 6 | 0.826 | 3 | 0.819 | 3 | 0.812 | 3 | | |
| 7 | 0.861 | 3 | 0.854 | 3 | 0.846 | 3 | | |
| 8 | 0.881 | 2 | 0.874 | 2 | 0.866 | 2 | | |
| 9 | 0.893 | 2 | 0.886 | 2 | 0.878 | 2 | | |
| 10 | 0.901 | 1 | 0.893 | 1 | 0.886 | 1 | | |

a nondecreasing function of the state variable $x$: Greater energy reserves result in increased fitness.

(2) For any given level of energy reserves $x$, the probability of surviving from $t$ to $T$ decreases with decreasing $t$: The forager is less likely to survive a long period than a shorter period.

(These qualitative predictions make good sense—in fact they can easily be deduced logically from the model, without numerical solution. We can therefore think of (1) and (2) as checks that our program is correct.)

(3) Although the solution of the dynamic programming equation produces values for the fitness function, the actual values themselves are perhaps not as interesting as the optimal decisions for each value of $x$ and $t$. The behavioral decisions—that is, which patch to choose—depend on the current value of the state variable $x$, and also on the current time $t$. This dependence of decisions on the organism's current state is one of the most important features of the dynamic approach to behavioral modeling. It is our view that biological realism is thereby greatly enhanced, relative to behavioral theories that do not consider the effect of the organism's state. (In mathematical jargon, a strategy in which decisions depend on the value of a state variable is called a "feedback" strategy: The organism monitors its state, and uses this information to influence its behavior.)

(4) Behavior changes with increasing time to go, $T - t$. For example, when $t$ is close to $T$, so that the time to go is relatively small, the behavior is either to pick the safe patch (1) or the risky patch (3). Risky behavior is optimal when the forager is close to the critical value of $X$, and safe behavior is optimal when the forager is far from the critical value. Next, note that as $t$ decreases further from $T$, choosing the less risky patch (2) sometimes becomes the optimal decision. This decision balances the risk of predation and future risks of starvation. Note too that the optimal decision may be different from the "myopic" decision based on $t = T - 1$, which is the decision if the organism were concerned only with surviving for one period. In the myopic case, choosing the moderately risky patch never appears as an optimal decision, but in the truly dynamic model it does.

(5) As the time to go increases, the behavioral decisions become relatively insensitive to the value of $t$ and depend essentially only upon the value of $x$. This phenomenon is called *stationarity*. That is, when many periods remain in the life of a forager, the current behavioral decision should be relatively insensitive to the final time horizon. (Whether an animal faces 229 days or 230 days through which it must survive before reproducing should not affect the current behavioral decision very much, if at all.) Stationarity is helpful in analyzing more complex and realistic problems because it leads to simplification in the interpretations of the results.

As an exercise, we encourage the reader to experiment with different values of $T$. What do you expect will happen?

We will now propose two "computer experiments" which involve minor modifications of the code just developed. Such experiments are analogs of laboratory experiments and are used in the same way—to help develop intuition and understanding about the processes under consideration.

*Experiment 1:* In the same way that we perform controls in laboratory experiments, it is always a good idea to test computer programs by using parameter values in which the output can be predicted. For example, if the safe patch has a probability equal to 1 of finding sufficient food, then we would always expect the optimal behavioral decision to be choosing the safe patch. Also, if the patches with some chance of predation are easily ranked (e.g. the chance of predation is lower, the chance of finding food is greater, and the value of food found is greater in Patch 3 than Patch 2), then it should be easy to predict optimal behavioral decisions. We encourage the reader to try experimenting with such "trivial" parameter values as a means of checking that the code is indeed working correctly.

*Experiment 2:* An interesting computer experiment involves a situation in which the two risky patches have identical expected food intakes. That is, modify the patch parameters so that $\lambda_2 Y_2 = \lambda_3 Y_3$. How do the optimal behavioral decisions change? What would you expect?

We encourage the reader to perform such computer experiments for a definite reason. One can gain considerable insight into the behavior of organisms by experimenting with the computer programs. In particular, computer experiments can be used to suggest real experiments—and the appropriate quantities to measure in such experiments. Further computer experiments are suggested in the next section, and in Appendix 2.1.

## 2.5 Elaborations of the Simplest Model

In this section we will discuss two modifications of the model just analyzed: (1) a more general terminal fitness function, and (2) a per-period fitness increment. These elaborations are extremely important for the applications discussed in later chapters. Some further elaborations are described in Appendix 2.1 to this chapter; while still important, these elaborations are not essential to the sequel, and may be skipped at a first reading.

*General terminal fitness functions.* In the models analyzed thus far, it is assumed that fitness is equal to the probability of survival up to a certain time horizon $T$. This specification of fitness assumes that only survival is important to the organism and that all values of $X(T) > x_c$ are equally valuable in terms of subsequent reproduction. In many cases, however, differing values of $X(T)$ may have differing implications for the future of the organism. For example, if at time $T$ the organism enters a breeding period, larger values of $X(T)$ may result in higher numbers of offspring or improved survival of the parent or its offspring, or both. This suggests that we introduce a *terminal fitness function*, which we will denote by $\phi(x)$, with the interpretation that $\phi(x)$ is the future fitness of the organism associated with terminal value $X(T) = x$. We then equate lifetime fitness to the expected value of $\phi(X(T))$:

$$F(x, t, T) = \max E\{\phi(X(T)) \mid X(t) = x\}. \qquad (2.14)$$

The interpretation of $F(x, t, T)$ is thus that it is the maximum expected value of the terminal fitness function based on behavioral decisions taken between time $t$ and time $T$, when the value of the state variable at $t$ is $x$.*

* It is worth remarking that the definition (2.14) of lifetime fitness

We will now demonstrate the remarkable fact that $F(x,t,T)$ defined by Eq. (2.14) satisfies exactly the same dynamic programming equation (2.10) as before, but has a different end condition. In particular, from the above definition of $F(x,t,T)$ we see that

$$F(x,T,T) = \phi(x). \tag{2.15}$$

Now let us consider how the equation that $F(x,t,T)$ satisfies can be derived, again employing the paradigm of patch selection used to derive Eq. (2.10). Consider a situation in which the organism starts at period $t$ with state variable $X(t) = x$. If it chooses patch $i$, then it survives period $t$ with probability $(1 - \beta_i)$; and then with probability $\lambda_i$ the value of the state variable at the start of period $t+1$ is $x_i'$, given by Eq. (2.8); and with probability $1 - \lambda_i$ the value of the state variable at the start of period $t+1$ is $x_i''$. In exact analogy to the verbal argument leading to Eq. (2.10), we now reason as follows: If patch $i$ is chosen in period $t$, then the expected fitness starting at period $t$ with $X(t) = x$ equals $(1-\beta_i)$ times the average of the maximum expected fitness starting at period $t+1$ if food is found plus the maximum expected fitness starting at period $t+1$ if food is not found. Put into symbols, this becomes again

$$F(x,t,T) = \max_i (1-\beta_i)\{\lambda_i F(x_i', t+1, T) + (1-\lambda_i) F(x_i'', t+1, T)\} \tag{2.16}$$

and this equation is identical to Eq. (2.10).**

---

$F(x,t,T)$ includes our previous definition (2.1) as a special case. Namely, suppose that

$$\phi(x) = \begin{cases} 1 & \text{for } x > x_c \\ 0 & \text{for } x \le x_c \end{cases}$$

i.e. the terminal fitness is one for $x$ greater than the critical value, and zero otherwise. It can then be shown that $F(x,t,T)$ as defined in Eq. (2.14) is equal to $\max \Pr(X(T) > x_c \mid X(t) = x)$, i.e. the maximum probability of surviving from $t$ to $T$.

** Mathematically speaking, Eq. (2.16) uses the Law of Total Expectation (1.22) in place of the Law of Total Probability which was used in deriving Eq. (2.10). However, both derivations follow by "common sense" considerations—namely, by adding up the various possibilities.

*Experiment 3:* Modify your existing patch selection code to include a terminal fitness function such as

$$\phi(x) = (x - x_c)/(x + x_0) \qquad \text{for } x > x_c \tag{2.17}$$

where $x_0$ is a parameter. What effect does this change have on the optimal patch strategy when the time to go is small and when the time to go is large? How can these results be explained?

*Per-period fitness increments.* There are many situations in which not only is the fitness of an organism assessed at the end of some time interval of length $T$, but fitness also accrues during each period. An example is the oviposition behavior of insects encountering and laying eggs in hosts in each period (see Chapter 4). Such situations can also be included as elaborations of the basic patch model. In order to do this, we first introduce a per-period fitness increment function $\psi_i(x)$ defined as follows:

$$\begin{aligned}\psi_i(x) = {} & \text{fitness accruing to the organism during period } t \\ & \text{if patch } i \text{ is visited and } X(t) = x.\end{aligned} \tag{2.18}$$

The expected lifetime fitness function $F(x, t, T)$ is now the sum of the expected per-period fitness increments and the terminal fitness function. Thus, we now define

$$F(x, t, T) = \max_i E\{\sum_{s=t}^{T-1} \psi_i(X(s)) + \phi(X(T)) \mid X(t) = x\}. \tag{2.19}$$

In this equation, the "max" again refers to a maximum over behavioral decisions (patch choices in this case). As always, the patch choice $i$ in each period $s = t, t+1, \ldots, T-1$ will depend upon the forager's state $X(s)$ at the beginning of that period.

As before, when $t = T$ the only fitness to the organism is associated with the terminal fitness function. Thus

$$F(x, T, T) = \phi(x). \tag{2.20}$$

Next consider values of $t < T$. If the value of the state variable at time $t$ is $X(t) = x$ and patch $i$ is chosen, three events must be considered. For simplicity, we assume that these events occur in the order of reproduction, predation risk, foraging. Thus the organism immediately receives an increment in fitness $\psi_i(x)$. If it survives period $t$ in patch $i$, the organism starts period $t+1$ with state variable equal to either $x_i'$ (with probability $\lambda_i$) or $x_i''$ (with probability $1-\lambda_i$). The optimal behavioral decision is thus to choose the patch that maximizes the combination of immediate increment in fitness plus expected future increments, starting with the new value of the state variable at period $t+1$. Thus, the dynamic programming equation in this case is

$$F(x,t,T) = \max_i [\psi_i(x) + (1-\beta_i)\{\lambda_i F(x_i', t+1, T) + (1-\lambda_i)F(x_i'', t+1, T)\}]. \tag{2.21}$$

*Experiment 4:* Modify your patch selection code to include the case of a per-period fitness increment, such as

$$\psi_i(x) = r_i x.$$

How does the optimal patch strategy change when this per-period fitness increment is used?

## REPRODUCTIVE VALUE

The fitness function defined in Eq. (2.19) is closely related to the concept of *reproductive value* introduced by R.A. Fisher (1930). Consider a population of organisms, and assume that the population is growing by the factor $r$ in each period. The appropriate fitness function in this case is given by

$$F(x,t,T) = \max_i E\Big\{ \sum_{s=t}^{T-1} r^{s-t}\psi_i(X(s)) + r^{T-t}\phi(X(T)) \mid X(t) = x \Big\}. \tag{2.22}$$

This function satisfies a dynamic programming equation analogous to (2.21), with additional factors involving $r$.

Now consider the special case in which there is no state variable $X(t)$, so that per-period reproduction is simply $\psi_i(s)$. Assume also that a fixed patch-selection strategy $i_1^*, i_2^*, \ldots, i_T^*$ is employed, so as to maximize fitness as defined by Eq. (2.22). Such behavior is completely nonresponsive, in the sense that behavior in period $t$ does not depend on the state of the organism or its environment. Define the survivorship functions

$$l(s) = (1 - \beta_{i_1^*})(1 - \beta_{i_2^*}) \cdots (1 - \beta_{i_s^*}).$$

Finally, assume that period $T$ is post-reproductive, i.e. $\phi(X(T)) = 0$. Then for $s \geq t$ we have

$$E\{\psi_i(s) \mid X(t) = x\} = \frac{\psi_{i_s^*}(s)l(s)}{l(t)}$$

(no reproduction occurs in period $s$ unless the animal survives). Hence Eq. (2.22) can be written in the form

$$F(x,t,T) = \sum_{s=t}^{T-1} r^{t-s} \frac{m(s)l(s)}{l(t)} \tag{2.23}$$

where $m(s) = \psi_{i_s^*}(s)$. Equation (2.23) agrees exactly with Fisher's (1930) definition of reproductive value.

To summarize, the lifetime fitness function $F(x,t,T)$ as defined in this chapter is a generalization of the classical notion of reproductive value *which allows behavior to be responsive to the current state of the organism* (as influenced by its past history), *and of the environment*. For simplicity, we treat only the case of stationary populations ($r = 1$), but the dynamic modeling procedure works equally well for growing populations. (In Chapter 8 we discuss a modification of the definition of lifetime fitness which may be appropriate for fluctuating environments.)

The relationship between the theory of life-history strategies and dynamic programming has been discussed by Leon (1976), Goodman (1982), and Schaffer (1983). These authors, however, do not stress the fact that the state variable approach provides a far-reaching *extension* of classical life history theory, allowing decisions to be responsive to unpredictable events that occur during the organism's life span. This flexibility is clearly essential in any analysis of behavior, as that term is usually understood.

## 2.6 Discussion

In this chapter we have outlined the dynamic framework for behavioral modeling in a particularly simple setting. We have indicated the flexibility of the approach by considering various simple modifications of the basic model. Several further elaborations of the model are discussed in Appendix 2.1. The remaining appendices discuss other aspects of the dynamic modeling approach. More advanced topics are discussed in Part III.

We wish to emphasize that the modeling framework introduced in this chapter can be adapted to a wide variety of behavioral phenomena. The structure of the model will change so as to reflect what is known (and considered important) about the biology. Likewise, the mathematical form of the dynamic programming equation may also change, depending on the particular situation. Many examples of these changes appear in the applications discussed in Part II.

## Appendix 2.1 Further Elaborations of the Patch Selection Paradigm

In this appendix we discuss several modifications and extensions of the basic patch selection model. These elaborations illustrate the flexibility of the basic paradigm. Similar elaborations will be encountered in many of the applications discussed in Part II.

### 2.1.1 Alternative Constraints

In cases where the state variable $X(t)$ represents energy reserves or gut capacity, constraints of the form

$$0 \leq X(t) \leq C$$

are natural. In other cases, however, it may be more realistic to assume no upper constraint on $X(t)$. An example would be the amount of food hoarded by a bird or a rodent.

*Experiment 5:* Modify your original patch selection code to include no capacity constraint. (Since the computer can only handle vectors of fixed dimension, this means that you must change $C$ in your code to a sufficiently large value, so that $X$ will never exceed $C$.) How does this change affect the results?

Another possible constraint would involve a maximum rate at which the state variable can increase:

$$X(t+1) - X(t) \leq R.$$

*Experiment 6:* Modify your original code to include a rate constraint. Hint: what effect will this have on the effective food size $Y_i$?

### 2.1.2 Variable Handling Times

The basic patch selection model characterizes patches in terms of the chance of predation, the chance of finding food, the cost of foraging in the patch, and the value of the food found. Another possible variable for characterizing the patch is the handling time that it takes to deal with a food item from the $i$th patch. Thus, let

$$\tau_i = \text{time required to handle a food item from the } i\text{th patch.} \tag{2.24}$$

Previously, we have implicitly assumed that all of the $\tau_i = 1$, but now we will allow $\tau_i > 1$.

If the forager encounters food $Y_i$ in the $i$th patch, the food will be consumed (subject to the capacity constraint) and $\tau_i$ units of time will be used up. This means that the basic dynamic programming equation (2.10) now becomes

$$\begin{aligned} F(x,t,T) = \max_i [&(1-\beta_i)^{\tau_i} \lambda_i F(x_i', t+\tau_i, T) \\ &+ (1-\beta_i)(1-\lambda_i) F(x_i'', t+1, T)] \end{aligned} \tag{2.25}$$

where now

$$x_i' = \text{chop}(x - \tau_i \alpha_i + Y_i; x_c, C). \tag{2.26}$$

An additional complication arises if $t + \tau_i > T$, which would mean that the animal encounters more food than it can eat in the time remaining. A simple way to finesse this possibility is to assume that the animal eats all remaining food in the terminal period (up to capacity). In this case, we just replace $t + \tau_i$ in Eq. (2.25) by $\min(t + \tau_i, T)$. Other methods for modeling the consumption of excess food at the terminal horizon can be imagined; details are left to the reader. Note also that Eq. (2.25) assumes that predation risk $\beta_i$ continues to apply for each of the $\tau_i$ periods required to consume the food item $Y_i$.

In order to implement Eq. (2.25), the algorithm of Section 2.4 must be expanded, since the computation of $F(x,t,T)$ now involves the values of $F(x, t + \tau_i, T)$ for various $\tau_i \geq 1$. If $\hat{\tau}$ denotes the largest of the $\tau_i$, then the computer code must keep track of $F(x, t', T)$ for $t' = t+1, \ldots, t+\hat{\tau}$. This is done by introducing an array $F1(x, \tau)$ which will correspond to the values of $F(x, t+\tau, T)$ for $0 \leq x \leq C$ and $1 \leq \tau \leq \hat{\tau}$. Initially, one sets $F1(x, 1) = \phi(x)$ and $F1(x, \tau) = 0$ for $\tau > 1$. The vector $F0(x)$ represents $F(x, t, T)$ as before.

Step 4 of the algorithm described earlier required updating of the array $F1(x)$ after $F0(x)$ had been computed. This step now becomes:

*Step* $4'$. Cycle over $\tau$ from $\hat{\tau}$ down to 2: for each value of $x$ replace $F1(x, \tau)$ by $F1(x, \tau - 1)$. Finally replace $F1(x, 1)$ by $F0(x)$. (What goes wrong if this step is carried out in the reverse order in $\tau$?)

*Experiment 7:* Modify the existing patch selection code to study the problem with variable handling times.

### 2.1.3 A Diet Selection Model

In order to show the versatility of the dynamic state variable approach, we will reinterpret the patch selection problem so that it can be viewed as a diet selection problem. "Optimal diet selection" is a theory concerned with the ways that predators should select prey; it has been part of foraging theory since the inception of the subject. Useful surveys are given by Pyke (1984), Pyke et al. (1977), and Stephens et al. (1986). A dynamic treatment of the diet selection problem appears in Houston and McNamara (1985).

For the present purpose we now reinterpret "patch" to mean food type, so that $\lambda_i$, $\alpha_i$, $\beta_i$, $\tau_i$, and $Y_i$ are now interpreted respectively as the probability of encountering food of type $i$ during period $t$, the cost of handling food of type $i$, the probability that a prey item of type $i$ can kill the foraging organism (usually we would set $\beta_i = 0$, but there may be instances in which $\beta_i$ would be nonzero), the handling time of food of type $i$, and the energetic value of food of type $i$. We again work with the expected lifetime fitness $F(x,t,T)$ defined in terms of an associated terminal fitness $\phi(x)$. Regarding the behavioral decision, we will assume that a food item may be either consumed or rejected. If a food item is encountered and rejected by the forager then the time variable is incremented from $t$ to $t+1$, whereas if the a food item of type $i$ is accepted by the forager, the time variable is incremented from $t$ to $t+\tau_i$. We introduce a basic metabolic rate $\alpha_0$ per period and assume that $\alpha_i = \tau_i\alpha_0$. Finally, we define a probability

$$\lambda_0 = 1 - \Sigma\lambda_i = \text{probability of not discovering any food item during period } t. \tag{2.27}$$

We encourage readers to try to derive the dynamic programming equation for the diet selection problem for themselves before reading any further. (There are some slight changes to the previous version.)

With this formulation, the dynamic programming equation for $F(x,t,T)$ is

$$\begin{aligned} F(x,t,T) = {} & \lambda_0 F(x_0'',t+1,T) \\ & + \Sigma\lambda_i \max\{F(x_0'',t+1,T); (1-\beta_i)F(x_i',t+\tau_i,T)\} \end{aligned} \tag{2.28}$$

where

$$\begin{aligned} x_i' &= \text{chop}(x - \tau_i\alpha_0 + Y_i; x_c, C) \\ x_0'' &= \text{chop}(x - \alpha_0; x_c, C). \end{aligned} \tag{2.29}$$

The first expression following the max represents the expected fitness if a prey item of type $i$ is found and rejected, and the second term represents the expected fitness if the prey item is accepted. Notice that the decision in this example is assumed to be made after the prey item is encountered. This accounts for the position of the max expression in Eq. (2.28).

*Experiment 8:* Modify your patch selection code to analyze the diet selection problem for three food types. In particular, address these two questions:

(a) When would the organism exhibit partial preferences, in that it will sometimes accept the $i$th food type and sometimes reject it?

(b) Are there any situations in which the forager will accept food types which provide an energy intake lower than the average over the environment? (Those readers familiar with diet selection "theorems" may wish to compare the results of the dynamic model with the usual theorems; see Houston and McNamara 1985 for a detailed analysis.)

### 2.1.4 A Model with "Fat Reserves" and "Gut Contents"

We next consider a model with two state variables: gut contents $G(t)$ and fat reserves $W(t)$ characterizing the state of the forager at time $t$. We assume that when forage is encountered, the gut contents are first increased by the forage consumed. Gut contents are used to meet metabolic requirements, and any excess is then converted into fat reserves. If gut contents are insufficient to meet metabolic needs then fat reserves are used instead. Figure 2.3 schematically illustrates our concept of the relationship between food found, gut contents, metabolic needs, and fat reserves.

To begin, assume first that at the end of each period the gut is cleared. This assumption might make sense if the period were a day, for example, so that gut clearance would occur overnight. In such a case we can really ignore the state of the gut contents and work with fat reserves $W(t)$ as a single variable. Before introducing the model, it is helpful to introduce one more mathematical symbol:

$$(z)_+ = \begin{cases} z & \text{if } z \geq 0 \\ 0 & \text{if } z < 0. \end{cases} \tag{2.30}$$

Suppose that the forager chooses to forage in patch $i$. We then assume that with probability $\lambda_i$ the dynamics of $W(t)$ are given by

$$W(t+1) = W(t) + \gamma(Y_i - \alpha_i)_+ - \delta(\alpha_i - Y_i)_+ . \tag{2.31}$$

where $\gamma$ and $\delta$ are, respectively, the conversion factor with which excess gut capacity is converted to fat reserves and the conversion factor for fat reserves to metabolic energy needs. Note that Eq. (2.31) is merely a restatement, in mathematical symbols, of

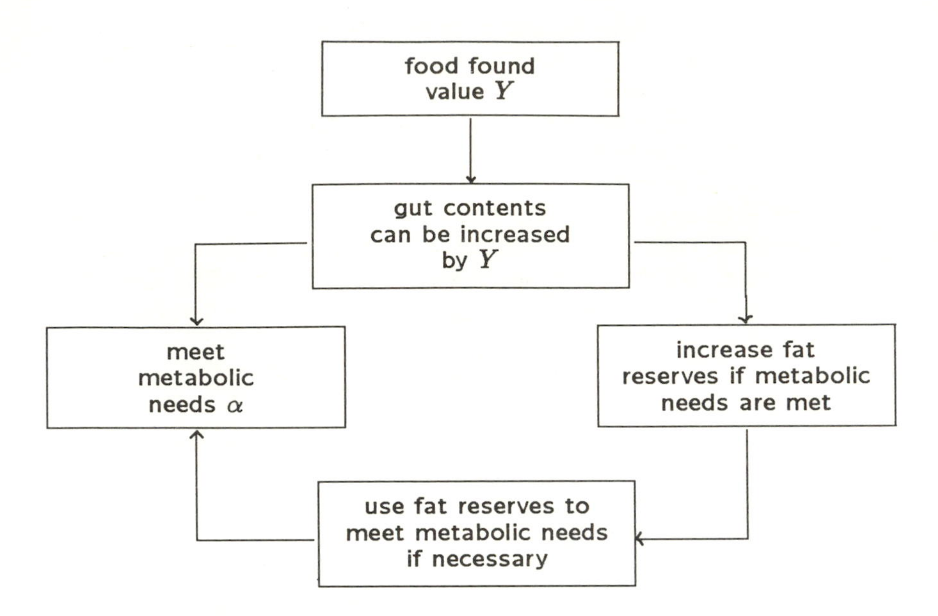


Figure 2.3 A simple model having gut capacity and fat reserves as state variables. If food is found, the gut contents are increased by $Y$, up to the capacity of the gut. The gut contents are then used to meet basic metabolic needs of the organism and excess gut contents can be converted to fat reserves. If gut contents are insufficient to meet the metabolic needs, fat reserves are converted to energy.

the assumption about the use of food and gut clearance. In particular, only one of the two terms of the form $()_+$ in Eq. (2.31) will be nonzero. If food eaten exceeds metabolic needs, then the excess is used to increase fat reserves; otherwise fat reserves are used up for metabolic needs. With probability $1-\lambda_i$ the dynamics of $W(t)$ are

$$W(t+1) = W(t) - \delta\alpha_i. \tag{2.32}$$

We now introduce a critical value of fat reserves, $W_c$, a terminal fitness function $\phi(W(T))$, and a lifetime fitness function $F(w,t,T)$ corresponding to the maximum expected fitness at time $T$, given that $W(t) = w$. The patch selection problem can then be solved in a fashion identical to before. We encourage the reader to modify the existing patch selection code and experiment with the dynamics (2.31) to see how behavioral decisions change. (Since $\gamma$ and $\delta$ will not have integer values, it will not be possible to restrict the values of the state variable $W(t)$ to integers. Hence some form of interpolation will be required in the code for the dynamic programming equation. The method of interpolation is described in Section 8.3.)

A more complicated model involves both state variables, fat reserves $W(t)$, and gut contents $G(t)$. There are many different assumptions that can be made concerning the dynamics of these variables; one example is the following. Assume that in each period a fraction $\kappa$ of gut contents at the start of that period in excess of metabolic cost is converted to fat reserves, but that food discovered during period $t$ cannot be converted to fat reserves until period $t+1$. Also assume that if the gut contents at the start of period $t$ are insufficient to meet the metabolic needs, then fat reserves are used. With these assumptions, the following dynamics characterize $G(t)$ and $W(t)$. Assume that patch $i$ is visited by the organism. Then with probability $\lambda_i$

$$\begin{aligned} G(t+1) &= (1-\kappa)(G(t)-\alpha_i)_+ + Y_i \\ W(t+1) &= W(t) + \kappa(G(t)-\alpha_i)_+ - \delta(\alpha_i - G(t))_+ \end{aligned} \tag{2.33}$$

and with probability $1-\lambda_i$ the dynamics are as in Eq. (2.33), except that $Y_i$ does not appear in the first equation.

The fitness function is now $F(g,w,t,T)$, defined as the maximum expected lifetime fitness associated with behavior between time $t$ and $T$, given that $G(t) = g$ and $W(t) = w$. It is still reasonable to assume that the organism dies if the fat reserves fall below the critical value. With these assumptions, the dynamic programming equation associated with the patch selection problem for these dynamics can be written down. The interested reader may wish to experiment, comparing results for models with and without fat reserves as a second state variable.

### 2.1.5 Sequential Coupling

The basic patch selection model is time-homogeneous, in the sense that the parameters $\alpha$, $\beta$, $\lambda$, $Y$ all remain constant over the entire time horizon. In reality these parameters will change over time, except possibly for relatively short time intervals. For example, a forager may face alternating days and nights. On a larger time scale, an organism may face alternating breeding and nonbreeding seasons. Ontogenetic development provides a third example.

It is still possible to use the basic dynamic framework when parameter values are time-dependent—one simply keeps track of the time dependence in the model and in the computer. However, an even simpler approach, which we call sequential coupling, can be used to break a time-dependent modeling problem into a sequence of time-homogeneous problems. Each of these simpler problems can be modeled, and the results interpreted, in a straightforward fashion as in the examples already discussed. Finally, the simpler models can be sequentially linked, or coupled. Several examples of this procedure will occur in the applications in Part II.

Let us illustrate the procedure of sequential coupling in the setting of the patch selection model. We consider a forager that makes a number of patch choices throughout the day, and then sleeps through the night. The process is repeated for $N$ days and nights. (A model of this sort is employed by McNamara et al. 1987.)

First we divide the day into $T$ periods of equal length, indexed by $t = 1, 2, \ldots, T$. Then let $X(t, d)$ denote the forager's state at the beginning of period $t$ on day $d$, where $d = 1, 2, \ldots, N$.

The dynamics of the state variable within a given day remain exactly the same as before. Overnight, however, the state variable is decremented by the nighttime metabolic cost $\alpha_n$. This can be expressed as:

$$X(1, d+1) = \begin{cases} X(T, d) - \alpha_n & \text{if } X(T, d) > \alpha_n \\ 0 & \text{if } X(T, d) \leq \alpha_n. \end{cases} \tag{2.34}$$

(We ignore the possibility of predation overnight, but this could easily be included.) Finally, define lifetime fitness as

$$F(x, t, T, d, N) = \max E\{\phi(X(T, N)) \mid X(t, d) = x\}. \tag{2.35}$$

The dynamic programming algorithm now proceeds as follows. First we consider the final day, $d = N$. We have

$$F(x, T, T, N, N) = \phi(x). \tag{2.36}$$

The dynamic programming equation on day $N$ is identical to Eq. (2.10):

$$\begin{aligned} F(x, t, T, N, N) = \max_i (1 - \beta_i)\{&\lambda_i F(x_i', t+1, T, N, N) \\ &+ (1 - \lambda_i) F(x_i'', t+1, T, N, N)\} \end{aligned} \tag{2.37}$$

where $x_i'$ and $x_i''$ are as before. This equation can be solved using the same code as in the original model.

Next comes the coupling step. Let us write

$$\phi_{N-1}(x) = \begin{cases} F(x - \alpha_n, 1, T, N, N) & \text{if } x > \alpha_n \\ 0 & \text{if } x \leq \alpha_n. \end{cases} \tag{2.38}$$

Note that the right-hand expression is already known, from solving the problem for day $N$. The important thing is that $\phi_{N-1}(x)$ then serves as the terminal fitness function for the problem on day $N-1$. Why? Because terminal fitness always represents future expected reproduction, after the terminal period. For day $N-1$ this is, also by definition, the same as the right side of (2.38).

The above coupling process can be continued for days $N-1, N-2, \ldots, 1$. For each day we use the same dynamic programming equation (already coded)

$$\begin{aligned} F(x,t,T,d,N) = \max_i (1-\beta_i)\{\lambda_i F(x_i', t+1, T, d, N) \\ + (1-\lambda_i) F(x_i'', t+1, T, d, N)\} \end{aligned} \tag{2.39}$$

and then we couple day $d$ to day $d+1$ by using the terminal fitness function for day $d$ as:

$$\phi_d(x) = \begin{cases} F(x - \alpha_n, 1, T, d+1, N) & \text{if } x > \alpha_n \\ 0 & \text{if } x \leq \alpha_n. \end{cases} \tag{2.40}$$

*Experiment 9:* Modify your original code to allow for $N = 5$ days and nights; assume that $\alpha_n = 5$, but in this case take $x_c = 0$. Is the optimal behavior any different on day one from day five?

### 2.1.6 Uncertain Final Time

The terminal time $T$ may correspond to the transition from one season to the next. If so, then $T$ itself may be a random variable, and the organism is faced with making decisions on the basis of an uncertain terminal time $T$. This possibility is easily incorporated into our basic model, as follows.

Let us first introduce the notation $T_f$ for the random terminal time, reserving the symbol $T$ for the largest possible value of $T_f$. Let $p(t)$ be the distribution of $T_f$:

$$p(t) = \Pr(T_f = t). \tag{2.41}$$

In general, terminal fitness at $t = T_f$ may depend on the state $X(T_f)$ and on $T_f$:

$$\text{terminal fitness} = \phi(X(T_f), T_f). \tag{2.42}$$

We consider only the case of no fitness increments, and define expected lifetime fitness for $t \leq T_f$ by

$$F(x,t,T) = \max E\{\phi(X(T_f), T_f) \mid t \leq T_f, X(t) = x\}. \tag{2.43}$$

The end condition is then

$$F(x,T,T) = \phi(x,T). \tag{2.44}$$

In order to write down the dynamic programming equation we need the conditional probabilities

$$p_f(t) = \Pr(T_f = t \mid T_f \geq t). \tag{2.45}$$

The method for computing $p_f(t)$ from $p(t)$ will be given later. The dynamic programming equation is

$$\begin{aligned} F(x,t,T) = {} & p_f(t)\phi(x,t) \\ & + (1 - p_f(t)) \max_i (1-\beta_i)[\lambda_i F(x_i', t+1, T) \\ & + (1-\lambda_i) F(x_i'', t+1, T)]. \end{aligned} \tag{2.46}$$

This is easily explained: with probability $p_f(t)$ the current period will be the final period, and fitness will equal $\phi(x,t)$. With probability $(1 - p_f(t))$ the current period is not final, so that fitness is obtained by optimal patch selection.

It is reassuring to note that Eq. (2.46) reduces to the basic dynamic programming equation (2.10) in the case of a nonrandom terminal time $T$. For in this case we have $p_f(t) = 0$ for $t < T$ and $p_f(T) = 1$.

The conditional probabilities in Eq. (2.45) are given by

$$p_f(t) = \frac{p(t)}{\sum_{s=t}^{T} p(s)}. \tag{2.47}$$

This equation follows immediately from Bayes' formula,

$$\Pr(A|B) = \frac{\Pr(B|A)\Pr(A)}{\Pr(B)}. \tag{2.48}$$

Bayes' formula follows immediately from the definition (1.7) of conditional probability:

$$\Pr(A|B) = \Pr(A \cap B)/\Pr(B) = \Pr(B|A)\Pr(A)/\Pr(B).$$

To obtain (2.47), let $A$ be the event $T_f = t$ and $B$ the event $T_f \geq t$. Then $\Pr(B|A) = 1$, and by the Law of Total Probability (1.11)

$$\begin{aligned}\Pr(B) &= \sum_s \Pr(T_f \geq t \mid T_f = s)\Pr(T_f = s) \\ &= \sum_{s=t}^{T} \Pr(T_f = s) = \sum_{s=t}^{T} p(s).\end{aligned}$$

*Experiment 10:* Modify your patch selection code to treat the case in which $T = 15$ with probability $1/4$ and $T = 20$ with probability $3/4$.

## Appendix 2.2
## Lifetime Fitness and Utility

Utility theory is an approach employed by economists to model decision making under uncertainty. The concept continues to be widely used, in spite of persistent controversy regarding its predictive value (see e.g. Kahneman and Tversky 1979, Schoemaker 1982, Heiner 1983, Machina 1987).

Ecologists have also used utility theory to model the behavior of organisms under uncertainty (e.g. Caraco et al. 1980, Rubenstein 1982, Krebs and McCleery 1984). It is not always clear, however, exactly what biologically meaningful entity is to be represented by utility, although in the words of Krebs and McCleery, "In a biological context... utility is taken to represent evolutionary fitness in some way." In this appendix we show that the lifetime fitness function $F(x, t, T)$ has the characteristics of a utility function. Thus in a dynamic model one can always work directly with

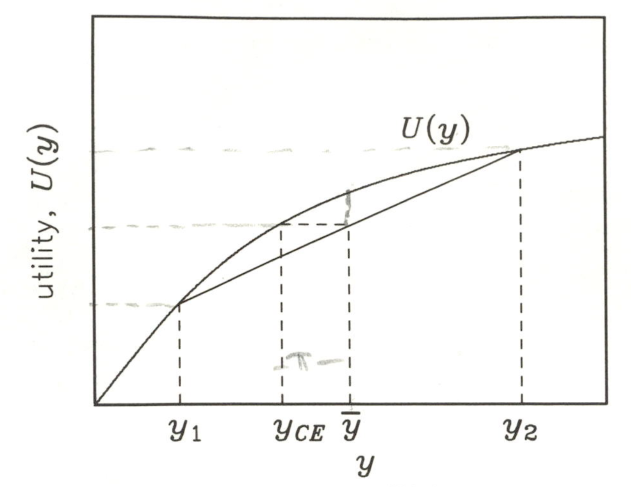


Figure 2.4 A concave utility function $U(y)$. The decision maker is "risk-averse," and has a positive "risk premium" $\pi = \bar{y} - y_{CE}$.

lifetime fitness and there is no need to introduce an extraneous concept such as utility. In simpler models the utility concept may retain some heuristic value.

Standard utility theory provides a preference ordering among "lotteries" of the form $(y_i, p_i)$, where $p_i$ denotes the probability of receiving the payoff $y_i$. The expected value of such a lottery is $\bar{y} = \Sigma p_i y_i$, but it is well known that decision makers do not always rank lotteries by their expected values.

On the basis of five axioms, postulated to describe rational decision making under uncertainty, von Neumann and Morgenstern (1947) demonstrated the existence of a utility function $U(x)$, defined over the decision maker's wealth $x$, with the property that the preference order of lotteries is determined by expected utility. The expected utility of the lottery $Y = (y_i, p_i)$, in which $y_i$ denotes an increment to wealth, is given by

$$E\{U(Y)\} = \Sigma p_i U(x + y_i). \tag{2.49}$$

The lottery $(y_i, p_i)$ is preferred to $(z_j, q_j)$ if and only if $E\{U(Y)\} > E\{U(Z)\}$. The utility function $U$ is unique to within an affine transformation $aU + b$.

The decision maker's attitude about risk is reflected in the shape of his utility function. First consider the case of a concave utility

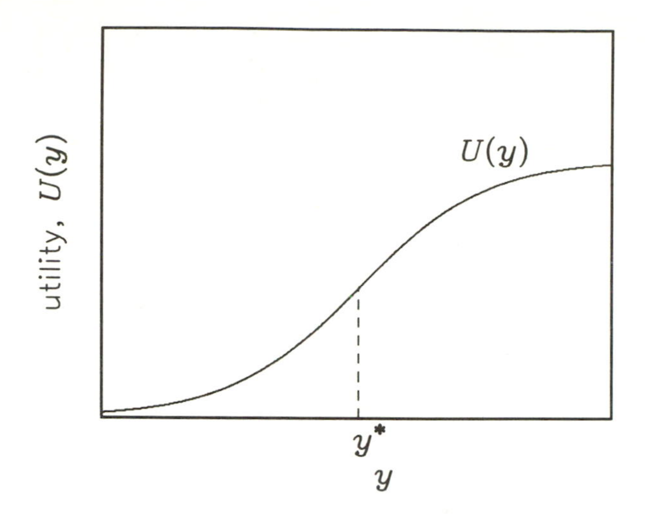


Figure 2.5 A convex–concave utility function $U(y)$; the decision maker is risk preferring for $y < y^*$ and risk-averse for $y > y^*$.

function (Figure 2.4). Assume that wealth $x$ is fixed; for simplicity we let $x = 0$. Suppose for the sake of illustration that $y_1, y_2$ are the possible payoffs of a lottery, with $p_1 = p_2 = 0.5$. Then $\bar{y} = 0.5(y_1 + y_2)$. The payoff $y_{CE}$ for which

$$U(y_{CE}) = E\{U(y)\} = \Sigma p_i U(y_i) \tag{2.50}$$

is called the *certainty equivalent* of the lottery; the decision maker is indifferent between playing the lottery and receiving the payment $y_{CE}$. The difference

$$\pi = \bar{y} - y_{CE} \tag{2.51}$$

is called the "risk premium" associated with the lottery.

When $U(x)$ is concave, as in Figure 2.4, we have $y_{CE} < \bar{y}$ (so that $\pi > 0$). In this case the value of the lottery to the decision maker is less than its expected value. Such a decision maker is said to be *risk-averse*. Conversely, a decision maker with a convex utility function is said to be *risk-prone*, or risk-preferring.

Many people enjoy gambling, but not for unduly large amounts of money. Such behavior displays risk preference for small bets and risk aversion for large ones. Behavior of this kind corresponds to a convex–concave utility function, as shown in Figure 2.5.

To summarize, the main feature of utility theory is the existence of a utility function $U(y)$ with the property that risky prospects, or "lotteries," are ranked according to their expected utility $E\{U(Y)\}$. Given a choice of one of several such prospects $Y_j$, a utility maximizer will select that prospect which maximizes $E\{U(Y_j)\}$.

Now let us consider the lifetime fitness function $F(x,t,T)$. The variable $x$, which represents the organism's current state, plays the same role as wealth in utility theory. The dynamic programming equation (2.10) can be expressed in the succinct form

$$F(x,t,T) = \max_i E\{F(x_i, t+1, T)\} \tag{2.52}$$

where $x_i = X(t+1)$ is the random value of the state variable in period $t+1$, given that patch $i$ is chosen. The random variable plays the same role as the lottery $Y$ in utility theory. In other words, the organism ranks the available activities (patches) according to their expected fitness. In this precise sense, a fitness optimizing organism behaves as a utility maximizer, where utility is identified with lifetime fitness. In the dynamic framework the "utility" function emerges from the model itself, and does not have to be invoked in an ad hoc fashion.

## Appendix 2.3
## Behavioral Observations and Forward Iteration

The methodology developed in Sections 2.3–2.5 allows us to compute the maximum probability of survival (or maximum expected lifetime fitness) and the optimal patch decisions. The optimal decision in period $t$ is a function of the current state of the forager, $X(t) = x$. We wish now to ask the question, what does the model predict about actual observations? In particular, what about the sequence of decisions taken by an individual organism? Also, what can be said about observations of a population of similar organisms?

First, since we are modeling a stochastic process, the model cannot predict the future state, or the actual sequence of decisions, of any individual. If an individual's state can be monitored over time somehow, then the model does predict behavior as a function of state, and this can be tested. If individual states cannot be monitored, then we are reduced to testing predictions at the population level. The purpose of this appendix is to show how state variable modeling can be used to make predictions about the fraction of organisms choosing a particular activity in a particular period.

First we assume that the dynamic programming equation has already been solved. Thus we know the optimal decision $c^*(x,t)$, which is the optimal choice of patch in period $t$ when $X(t) = x$. We now introduce the *state distribution vector* $P(x,t)$ defined as follows:

$$P(x,t) = \Pr(\text{organism has state } X(t) = x, \text{ given that organism is following the optimal behavioral decisions}). \tag{2.53}$$

We assume that an initial distribution is given. For example, if

$$P(x,0) = \begin{cases} 1 & \text{if } x = x_0 \\ 0 & \text{otherwise} \end{cases} \tag{2.54}$$

then the organism starts with the state variable equal to $x_0$ at time 0. Alternatively, one could set $P(x_i, 0) = p_i$, where $0 \leq p_i \leq 1$ and the $p_i$ sum to 1. In this case there is a distribution for the initial value of the state variable.

There are two interpretations of $P(x,t)$. The first is that it is the probability that an individual organism has $X(t) = x$. The second is that it is the fraction of organisms in a population with $X(t) = x$. In order to use this second interpretation, one must assume that all members of the population are following the same behavioral strategy and that the individuals encounter prey and predators independently with the same probability distribution.

The equation that $P(x,t)$ satisfies can be obtained by *forward iteration* (in contrast to the backward iteration used in the dynamic programming equations). In order to derive this equation, it is helpful to introduce the concept of a *transition density* which we denote by $w(x,t \mid z)$ and define in the following way:

$$w(x,t \mid z) = \Pr(\text{organism following the optimal behavioral decisions has } X(t) = x \mid X(t-1) = z) \tag{2.55}$$

The values of $w(x,t \mid z)$ can be calculated using the knowledge of $c^*(x,t)$—see below. Application of the Law of Total Probability then leads to the forward iteration equation

$$P(x,t) = \sum_z w(x,t \mid z)P(z,t-1). \tag{2.56}$$

This equation is then iterated forwards in time, beginning with the initial density $P(x,1)$.*

The next step is the calculation of the transition matrix $w(x,t \mid z)$. It is easiest to describe the development of this matrix in the form of an algorithm that could then be used in conjunction with the patch selection codes that we have described previously. Assume for simplicity that the critical value is $x_c = 0$.

**Algorithm for computation of $w(x,t \mid z)$:**

*Step 1.* Cycle over all values of $z$ from 0 to $C$.
*Step 2.* Given $z$, let $i^* = c^*(z,t-1)$; this is the patch chosen in period $t-1$.
*Step 3.* Put

$$w(x,t \mid z) = \begin{cases} (1-\beta_{i^*})\lambda_{i^*} & \text{for } x = \min(z - \alpha_{i^*} + Y_{i^*}, C) \\ (1-\beta_{i^*})(1-\lambda_{i^*}) & \\ & \text{for } x = z - \alpha_{i^*} \text{ if } z - \alpha_{i^*} > 0 \\ \beta_{i^*} & \text{for } x = 0 \text{ if } z - \alpha_{i^*} > 0 \\ \beta_{i^*} + (1-\beta_{i^*})(1-\lambda_{i^*}) & \\ & \text{for } x = 0 \text{ if } z - \alpha_{i^*} \le 0 \\ 0 & \text{for all other values of } x. \end{cases}$$

It should be clear how to modify this algorithm to cover alternative forms of the basic patch selection model.

Once $P(x,t)$ is known, we can compute the fraction of organisms choosing patch $i$ in period $t$ by simply adding up the probabilities for state variable values for which patch $i$ is optimal. Define

$$f_i(t) = \text{fraction of organisms (or probability of an organism) choosing patch } i \text{ during period } t. \tag{2.57}$$

* The more mathematically inclined reader will recognize that Eq. (2.56) describes a non-stationary Markov chain. The chain is stationary if $w(x,t \mid z)$ is independent of $t$.

To write $f_i(t)$ compactly, it helps to introduce one more notation, called the Kroenecker delta function:

$$\delta_{i,c^*(x,t)} = \begin{cases} 1 & \text{if } c^*(x,t) = i \\ 0 & \text{otherwise.} \end{cases} \tag{2.58}$$

Then the fraction of organisms choosing the $i$th patch is simply computed by

$$f_i(t) = \sum_x P(x,t)\delta_{i,c^*(x,t)}. \tag{2.59}$$

The fractions $f_i(t)$ are what one would most likely observe in a behavioral experiment.

> *Experiment 11:* Write out the $8 \times 8$ transition matrix $w(x, 2 \mid z)$ for the original patch selection model (see Table 2.1). Note that the transition matrix $w(x, t \mid z)$ is independent of $t$ (i.e. stationary) for $t \leq 12$ but changes for larger $t$.

## Appendix 2.4
## The Fitness of Suboptimal Strategies

When we solve the patch selection problem with an arbitrary terminal fitness function (so that Eq. (2.16) is the dynamic programming equation with Eq. (2.15) as the end condition), we obtain as part of the solution the set of optimal decisions. These are denoted by $c^*(x,t)$, representing the optimal patch to visit in period $t$, if the state variable at the start of period $t$ is $x$. In this appendix, we consider the fitness associated with suboptimal strategies. That is, suppose a strategy is specified by giving $c(x,t)$, representing which patch will be visited in period $t$, if the value of the state variable at the start of period $t$ is $x$. Thus, if $c(x,t) = c^*(x,t)$ for all $t$ and $x$, then $c(x,t)$ is the optimal strategy. We let $F_c(x,t,T)$ denote lifetime fitness associated with the strategy $c(x,t)$. Since no patch choice is involved in period $T$, we have

$$F_c(x,T,T) = \phi(x) \tag{2.60}$$

so that $F_c(x,t,T)$ and the optimal lifetime fitness $F(x,t,T)$ have the same terminal value. Unlike the optimal lifetime fitness, the suboptimal fitness $F_c(x,t,T)$ satisfies a simple iteration equation

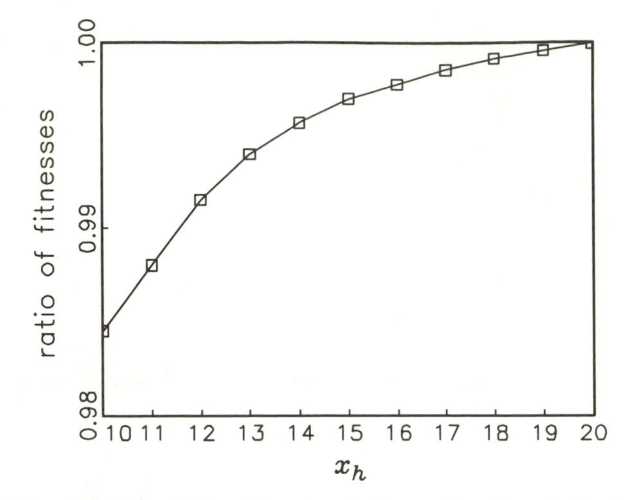


Figure 2.6 Ratio of the fitness of the suboptimal strategy $c(x,t)$ to the fitness of the optimal strategy for the basic patch selection problem. The suboptimal strategy is "hide" if $x > x_h$ and otherwise use the optimal patch choice.

instead of a dynamic programming equation. This iteration equation is

$$F_c(x,t,T) = (1-\beta_{c(x,t)})\{\lambda_{c(x,t)} F_c(x'_c, t+1, T) + (1-\lambda_{c(x,t)}) F_c(x''_c, t+1, T)\} \tag{2.61}$$

where $x'_c$ and $x''_c$ are given by Eq. (2.8) with $i = c(x,t)$.

The ratio

$$R_c(x) = F_c(x,1,T)/F(x,1,T) \tag{2.62}$$

provides a measure of the relative decrease in fitness caused by following the suboptimal strategy $c(x,t)$ instead of the optimal strategy $c^*(x,t)$. (This measure is similar in spirit to the concept of "canonical cost" introduced by McNamara and Houston 1986).

For example, consider the following definition of $c(x,t)$ for the basic patch selection problem with terminal fitness measured by probability of survival:

$$c(x,t) = \begin{cases} c^*(x,t) & \text{if } x < x_h \\ 1 & \text{if } x \geq x_h. \end{cases} \tag{2.63}$$

That is, the strategy $c(x,t)$ is to hide if the state variable exceeds a value $x_h$ and to behave optimally otherwise.

Figure 2.6 shows the value $R_c(20)$ for the patch parameters in the main text with $T = 30$ and $C = 20$. When $x_h$ is near the capacity $C$, the fitness obtained with the alternative strategy is nearly as large as that obtained from the optimal strategy (i.e. $R$ is nearly 1). As $x_h$ decreases, the suboptimal strategy becomes more and more inferior.

When the fitness obtained by using a suboptimal strategy is nearly as great as that obtained from the optimal strategy, the selection pressure for behaving exactly according to the prediction of the optimization model is relatively weak (Houston and McNamara 1986).

Given that the model is always a simplification of reality, it follows that observed behavior can be expected to depart to some extent from the model's exact predictions. Also, because of spatial or temporal inhomogeneities not encompassed by the model, observed behavior may be somewhat more variable than predicted. If the departure seems excessive, one should check the fitness implications by computing the ratio $R_c(x)$. If this ratio turns out to be low, then one has obtained evidence for the rejection—or perhaps modification—of the model. We emphasize that is not departures from predicted behavior as such, but departures from predicted maximum fitness, that provide such evidence.

*Experiment 12:* Modify your patch selection code so that, after solving the dynamic programming equation, you can compute $F_c(x, t, T)$ and $R_c(x)$ for alternative strategies $c(x, t)$. Try your code on some interesting (e.g. nonadaptive, $c(x, t) =$ constant) strategies.

# Addendum to Part I How to Write a Computer Program

Many of the students to whom we have taught the material in this book have had little or no programming experience. Our approach to them was "sink or swim"—we just recommended that they obtain an elementary book on programming (probably in BASIC) and do it! Many eventually overcame their fears and inexperience, and produced programs that worked. However, many students also pleaded with us to write a brief guide for absolute beginners. This is it.

First, don't allow yourself to be overwhelmed by the fact that the computer has many features that you don't understand. If you've never even touched a computer keyboard before, it would be worthwhile getting some sympathetic friend to show you a few simple things. If this is not possible, you'll have to check the beginning pages of your computer manual, which may be called "Operating System," "DOS," or some such thing. Remember that *nothing you can type on the keyboard can harm the computer*,* so don't hesitate to experiment.

We'll assume that you have access to a microcomputer, with the BASIC programming language.** Each make of computer seems to have its own peculiar version of BASIC, but they are all quite similar. There should be a BASIC handbook for your particular computer, which will tell you how to start your BASIC system, and which will also describe all the commands and programming features that are available. We suggest you keep this handbook at your side for reference. To begin with, you should read how to

* The only exception would be misuse of the **FORMAT** command while attempting to format a new floppy disk. A mistake here can destroy information stored on the computer's hard disk (if it has one). Don't worry about this until you get to using floppy disks for storage.

** We use BASIC here because it is widely available, and easy for beginners. Most programmers eventually graduate to a more sophisticated programming language such as FORTRAN or PASCAL. On the other hand, several advanced versions of BASIC are now available for microcomputers.

start BASIC, how to type and **RUN** a BASIC program. Soon you will also want to learn about commands such as **EDIT**, **SAVE**, **LOAD**, **LIST**. These commands, which handle your programs, will not be discussed here.

We recommend that you read this Addendum while seated at the keyboard. Type in the sample programs and try running them. Try out a few modifications—how does the computer respond to various kinds of errors?

*What is a program?* A computer program is just a list of instructions telling the computer to perform certain operations. A typical program will:

(1) input some data
(2) manipulate the data
(3) output some results.

Large commercial programs, sometimes called "software," still follow the same pattern. For example, a word processor first inputs data (what you type in), then manipulates the data (formats lines, paragraphs, etc.), and finally outputs the results (prints the document). Computers can manipulate numerical, alphabetical, or symbolic data of any kind, once the software is provided.

## *Program 1*

```
10  REM PROGRAM 1: COMPUTE X SQUARED
20  INPUT "X"; X
30  Y = X * X
40  PRINT X "SQUARED EQUALS" Y
50  END
```

Program 1 is a very simple example of a BASIC program. You can probably figure out what it does simply by reading it. BASIC is a "high-level language," in the sense that it looks much like plain English. Type Program 1 into your computer, and try running it.*

Note that each line is numbered. When you **RUN** a program, the computer performs the instructions in numerical order (unless

* Because of minor variations in individual BASIC packages, Program 1 as listed could conceivably fail to run on your computer. This would probably be signified by a **SYNTAX ERROR** message on the screen. If this happens, check with your BASIC handbook to see how your version uses certain instructions such as **INPUT** and **PRINT**.

there is an instruction telling the computer otherwise—see below). Let us inspect each of the instructions in Program 1.

```
10  REM PROGRAM 1: COMPUTE X SQUARED
```

The letters REM identify this as a *remark*. All remarks are completely ignored by the computer. Remarks are used to help the programmer remember what the program is about; they can be inserted anywhere. Including useful remarks is called documenting your program.

```
20  INPUT "X"; X
```

When the computer comes to this instruction, it types

```
X?
```

on the screen and awaits your response. When you type some number, say 7.22 (followed by RETURN or ENTER—see your computer handbook), the computer *inputs* the number as X. What actually happens is that the computer allocates a memory location for a number, names the location X for future reference, and stores your number 7.22 in location X.

The INPUT instruction is one of several ways that data can be entered into the computer. Some other methods of entering data will be described later.

```
30  Y = X * X
```

Now the computer manipulates the data stored in X. The symbol "*" means multiplication in BASIC, so that line 30 tells the computer to calculate X times X, or $X^2$, and to store the result in a new variable (memory location) called Y.

In general, any statement of the form*

Y = (something)

is an *assignment statement*. The computer first works out the value of the expression (something), then stores it in the variable Y. The expression (something) can be any expression that makes

* The statement can also be written as LET Y = (something).

sense to the computer, and uses values that are already stored in memory. However, the assigned variable Y must be a simple variable name—not an expression. For example the instruction

```
X * X = Y
```

is meaningless in BASIC, and would elicit a SYNTAX ERROR from the computer. Try it! (Type 30 X * X = Y RETURN. This replaces line 30 by the new line. Then type RUN.)

Note especially that the symbol " = " as used in BASIC does *not* mean mathematical equality. It means assignment. (Other languages may use different symbols for assignment.) The statement

```
X = X + 1
```

makes perfect sense in BASIC; it says "retrieve the number stored in X, add 1 to it, and store the result back in X." Statements like this are very often used in programming.

What happens if you use an assignment statement with an unspecified variable on the right-hand side? Try the following program (first type NEW to wipe out Program 1):

```
10  Y = Z + 2
20  PRINT Z, Y
```

How does the computer know what Z is? It doesn't! The result may depend on your particular computer, although probably it will assume that Z = 0.* Of course you would never deliberately have an unspecified variable in a program, but suppose you make a typing mistake. You will then have a "bug" in your program, but the computer may just go on merrily as if everything is okay. It can be very annoying to try to locate such typing bugs.

```
40  PRINT X "SQUARED EQUALS" Y
```

This is the output portion of Program 1. The message surrounded by quotation marks is called a STRING, and is printed literally. The values X and Y are printed in numerical format. See

* Your computer screen and most printers may print zero as Ø to distinguish it from the upper case letter O.

your handbook for formatting possibilities using PRINT and PRINT USING, and also LPRINT.

```
50  END
```

Signifies end of the program.

*Exercise 1:* Edit Program 1 to calculate the square root of X (the symbol SQR(X) is used by BASIC for calculating square roots). What happens when you input a negative value for X?

## VARIABLE NAMES

So far we have used single symbols like X,Y as variable names. However, BASIC allows whole strings, such as ENERGY or X5 to be used as variable names (but you mustn't use reserved words like INPUT or FOR); your BASIC handbook lists the conventions for naming variables.

## LOOPING

Instead of calculating a single square $x^2$, Program 2 calculates the sum $\sum_{x=1}^{N} x^2$. Note again the input data–manipulate–output sequence. The sum is computed by the FOR–NEXT loop in lines 40-60. First the variable SUM is initialized to SUM = 0 in line 30. Then the loop increments SUM by X * X = X$^2$ for X = 1 up to N. When the loop is exited, SUM equals $\sum_{x=1}^{N} x^2$, which is printed.

### *Program 2*

```
10  REM PROGRAM 2: SUM OF SQUARES
20  INPUT "UPPER LIMIT OF SUMMATION"; N
30  SUM = 0
40  FOR X = 1 TO N
50       SUM = SUM + X*X
60  NEXT X
70  PRINT "SUM OF" N " SQUARES EQUALS" SUM
80  END
```

Looping is one of the most useful programming features. In general, the effect of the loop

```
100  FOR INDEX = IO TO I1
110
 .
 .   (statements)
 .
190
200  NEXT INDEX
```

is transparent: the list of statements between the FOR statement and its corresponding NEXT statement is performed in sequence for each value of INDEX within the indicated range. (For ease of reading programs it is useful to indent the statements of the loop, as in Program 2.)

Read your handbook's description of the FOR...NEXT statement; there are several features, including STEP which has the index step different from 1. Note in particular that FOR...NEXT loops can be nested, an essential ingredient of many programs; see Program 4 below.

Some BASIC systems have alternative looping statements, such as WHILE...WEND or REPEAT...UNTIL, which can be very useful, but are not necessary. Check your handbook at your leisure.

## Decisions

Consider the expression (see Eq. 2.5)

$$\text{chop}(x; a, b) = \begin{cases} x & \text{if } a \leq x \leq b \\ b & \text{if } x > b \\ a & \text{if } x < a. \end{cases}$$

Here is a simple program segment that achieves this result; the value of $x$ has already been assigned earlier in the program.

```
 .
 .
 .
200  IF X < A THEN X = A
210  IF X > B THEN X = B
 .
 .
 .
```

Statement 200 does just what it says: if X is less than A then X is replaced by A. If X $\geq$ A then statement 200 does nothing. Statement 210 is similar. Thus the value that X had before statement 200 is replaced by chop(X;A,B) after 210 has been performed. (How would you modify this segment if you wanted to put Y = chop(X;A,B), leaving X unchanged?)

In general, the instruction

```
IF (a) THEN (b)
```

tells the computer to check whether (a) is true, and if so the computer carries out instruction (b). If (a) is false, nothing happens and control is passed to the next numbered statement in the program.

A useful variant is

```
IF (a) THEN (b) ELSE (c)
```

which is available in many BASIC packages. It should be clear how this works—if (a) is true then (b) is performed, while if (a) is false then (c) is performed. For example

```
IF X >= 0 THEN Y = SQR(X) ELSE Y = SQR(-X)
```

computes $Y = \sqrt{|X|}$ (note that $>=$ means "greater than or equal to"; check your manual for the various possibilities). Note that the ELSE part of the statement must occur on the same line as the IF...THEN part; you cannot start a new line number with ELSE.

Next, what would you do if you wanted the computer to perform a whole sequence of instructions if (a) is true, and a different sequence if (a) is false? Consider this segment:

```
200  IF (a) THEN GOTO 300
210
 .
 .   } instructions performed if (a) is false
 .
280
290  GOTO 400
300
 .
 .   } instructions performed if (a) is true
 .
390
400
```

Here, if (a) is true, the computer skips to line 300, so that the instructions 210 to 290 are not performed, but 300-390 are. Conversely, if (a) is false then the GOTO 300 instruction is not performed, so that instructions 210-290 *are* performed. Note that line 290 then causes the computer to skip lines 300-390.

The IF...THEN...(ELSE...) statement is extremely flexible; many practical computer programs could not be written without this feature.

## I/O

The acronym I/O in computerese stands for "input/output." Information can be fed into and out of the computer in various ways. For example, to enter data you can either (a) type it right into the program itself, or (b) enter it interactively from the keyboard. (You could also arrange the program to read data from a separate data file stored on disk, but this requires familiarity with file handling. Unless you have a lot of data this method is not recommended.)

The use of the INPUT statement to enter data interactively in BASIC has already been described. You should use this method *only* for data that will be changed almost every time you run the program; even then it's not essential to use INPUT, because program lines are so easy to EDIT in BASIC.

There are two simple methods to include data right in a program. For example, suppose your data consists of the parameter values

$$\beta = 0.3, \quad \alpha = 1.0, \quad \lambda = 0.05, \quad C = 50$$

Either of the following program segments will enter these parameters:

```
50  BETA = 0.3
60  ALPHA = 1.0
70  LAMBDA = 0.05
80  CAPACITY = 50
```

or

```
50  DATA 0.3, 1.0, 0.05, 50
60  READ BETA, ALPHA, LAMBDA, CAPACITY
```

The READ-DATA statements are always used together, although they can appear anywhere and in any order, in your program; see your handbook.

*Exercise 2:* What does Program 3 do? What would it do with a longer or shorter list of data (ending with −1)?

### *Program 3*

```
10  DATA 14, 28, 5, 42, 31, 29, 22, 40, -1
20  SUM = 0
30  I = 0
40  READ X
50  IF X < 0 THEN GOTO 90
60  SUM = SUM + X
70  I = I + 1
80  GOTO 40
90  PRINT "AVERAGE SCORE IS " SUM/I
100  END
```

*Exercise 3:* Extend Program 3 so that it will also print out the largest and smallest of the scores. (Assume that the scores in DATA are all between 0 and 50, say.) Try your program on the computer.

## BUILT-IN FUNCTIONS

BASIC has several useful "built-in" functions, including the trigonometric functions SIN, COS, TAN, and ATN (arc tangent), as well as the exponential EXP and natural logarithm LOG. You don't have to worry about how these functions are actually computed. (BASIC contains quite sophisticated, optimal algorithms which actually compute the functional values whenever called upon—it does *not* look them up in a table!) Other functions include the absolute value ABS and square root SQR functions. The result of a statement such as

```
100  Y = SQR( 1 + 2*COS(X) )
```

is self-evident.

Various other functions such as INT, SGN, and RND may also be available with your BASIC language. Your manual should contain a list of all available functions, and a description of what they accomplish.

By the way, the expression $a^x$ is written in BASIC as

```
A^X,
```

but other languages may use a different notation.

## ARITHMETIC OPERATIONS

Suppose you type the line

```
Y = A + B * C
```

How will the computer interpret this? It could be `(A+B) * C`, or it could be `A + (B*C)`. In fact BASIC does the latter. If you have not already done so, you should read the section of your handbook about arithmetical (also logical and relational) operators, and order of execution. For ease of reading your programs, it is a good idea to include parentheses in expressions that might be ambiguous, even if BASIC has a definite convention. For example

```
A / B * C
```

could equally mean `(A/B) * C` or `A / (B*C)`—use parentheses in your program to indicate which meaning is required.

## ARRAYS

An *array* is a block of numbers `A(I,J,K,...)`, where the numbers `I,J,K,...` are called *subscripts*. An array with only one subscript `I` is the same as a vector, while an array with two subscripts is a matrix. Arrays with more than two subscripts have no name, but BASIC allows up to 255 subscripts!

Every array in a BASIC program must be dimensioned with a `DIM` statement, such as

```
20  DIM FITNESS(20,10)
```

After this statement, the expression `FITNESS(I,J)` can be used as a variable name, anywhere in your program, for any value of $I = 0, 1, \ldots, 20$ and $J = 0, 1, \ldots, 10$. It can often be convenient to have the first value of the subscript be zero (of course this means that `FITNESS` is really a $21 \times 11$ matrix).*

* Some BASIC systems may not use 0 as a subscript, in which case $I = 1, \ldots, 20$ and $J = 1, \ldots, 10$ in this example.

READ-DATA statements are especially useful for entering an array of data values:

```
40  DIM J(5)
50  FOR I = 1 TO 5
60    READ J(I)
70  NEXT I
80  DATA 4.2, 0.22, 1.13, -6.1, 0.06
```

How you organize your data input is largely a matter of personal taste. Usually you will want to do all data input at the beginning of your programs (or in a separate subroutine—see below).

Program 4 uses nested loops to read the elements of a matrix A(I,J).

### *Program 4*

```
10  M = 3: N = 4: REM DIMENSIONS OF MATRIX A
20  DATA 5, 1, -2, 3
30  DATA 1, 0, 3, -1
40  DATA 2, -2, 0, 1
50  DIM A(M,N)
60  FOR I = 1 TO M
70     FOR J = 1 TO N
80       READ A(I,J)
90     NEXT J
100  NEXT I
```

Line 60 first fixes the value of I = 1. Then the loop in lines 70-90 reads A(1,1),...,A(1,N). Next, line 100 increments I to 2 and the next row A(2,1),...,A(2,N) is read, etc.

*Exercise 4:* Suppose N = M. Extend Program 4 to test whether the matrix A is symmetric (i.e. whether A[I,J] = A[J,I] for all I,J). Try your program to see if it works.

## SUBROUTINES

A computer program designed to perform useful tasks may turn out to be quite long—1,000 or even 10,000 lines of code are not unusual. Obviously such long programs can become very unwieldy—hard to write, and even harder to read. Various methods can be used to break long programs into shorter, largely independent pieces, thereby producing "structured" programs—see below.

One of the most important of these methods is the use of *subroutines.* Program 5 reads in a matrix $(a_{ij})$, then uses a subroutine to compute the row sums $s_i = \sum_{j=1}^{m} a_{ij}$ and print these out on the line printer.

### *Program 5*

```
10  M = 5: N = 8: REM DIMENSIONS OF MATRIX A
20  DATA ....
30  DIM A(M,N)
40  FOR I = 1 TO M
50     FOR J = 1 TO N
60       READ A(I,J)
70     NEXT J
80  NEXT I
90  LPRINT " ROW #           SUM"
100  LPRINT " -----------------"
110  FOR I = 1 TO M
120    GOSUB 1000
130    LPRINT I, SUM
140  NEXT I
150  LPRINT " -----------------"
160  END
1000  REM SUBROUTINE FOR ROW SUMS
1010  SUM = 0
1020  FOR J = 1 TO N
1030    SUM = SUM + A(I,J)
1040  NEXT J
1050  RETURN
```

The loop in lines 110-140 computes the row sums by referring to the subroutine beginning on line 1000, and then prints out the results. The statement GOSUB 1000 instructs the computer to transfer control to line 1000, where the subroutine begins. The RETURN statement in line 1050 at the end of the subroutine transfers control back to the main program at the line following the point of departure, i.e. line 130. Note that the subroutine is "called" once for each value of I in the loop.

*Exercise 5:* Extend Program 5 so as to compute and print the maximum element in each row, using a second subroutine.

One subroutine can call another subroutine, and so on. Note that subroutines must end with a `RETURN` statement, and main programs must end with an `END` statement. Failure to include these statements will lead to bizarre results.

## Other Features

Your BASIC package also contains commands and functions pertaining to the manipulation of alphabetical text ("strings"), the handling of files, and perhaps graphics. We will not discuss these features here, however, as they are not needed for the purposes of this book.

## Debugging

You now know enough to be able to write arbitrarily long, complicated programs for scientific purposes. Unfortunately, it is exceedingly easy to make mistakes. Any kind of mistake in a computer program is known affectionately as a "bug." It's just amazing how often you have to spend three days debugging a program that only took three hours to write in the first place. Strong men—even strong women—have been known to weep openly while debugging programs.

There are many species of bugs, but here are some of the more common genera:

(1) The typo: `SUN` instead of `SUM`. One hint to avoid typo bugs is to use recognizable words as variable names. `ENERGY` is a more readable name than `XEN`, or `X2`.

(2) The syntax error: `2(X+1)`; should have been `2*(X+1)`. This will be flagged when you try to run your program. Don't kick the computer when it snidely says `SYNTAX ERROR`—it's only trying to help.

(3) Logic bugs: these are errors in which the program performs differently from what you want. Logical errors usually result from too much haste in moving from the planning stage to coding the program. Because the program actually runs (but produces the wrong answer), logical bugs can be very difficult to locate. Sometimes you may not even know at first that the results ***are*** wrong.

The best way to avoid logical bugs is to adopt careful, rigorous programming habits. You should always verify the correctness of your program before running it. Finally, the use of a "structured programming" style will greatly improve your success rate, and also extend the usefulness of your programs by making them easily understandable. We will briefly discuss these three topics.

### Programming Habits

The following steps should be followed in writing any program:

1. Develop the necessary algorithms.
2. Write pseudocode which realizes the algorithms.
3. Code the pseudocode into an actual program.
4. Verify the program by hand (see below).
5. Test the program.

Clearly it doesn't make sense to try to write a program if you don't understand the algorithm you're going to use. *Pseudocode* is just a written list of the main steps in your program. Good, clear pseudocode, while it may only take a short time to write down, can save time and prevent errors in writing the program. Make your computer code follow your pseudocode as closely as is feasible.

### Verifying the Program

Once you have actually typed your program into the computer, it is an excellent idea to print a hard copy of it (use LLIST—check your manual), and then to verify it by hand. This might seem impossible (am *I* a computer?) but in fact it is quite straightforward, and will often eliminate logical bugs that could otherwise be very difficult to find. Also, verification will give you confidence that your program will do what it's supposed to.

Let us consider Program 3 again, with simplified DATA:

```
10  DATA 14, 28, -1
20  SUM = 0
30  I = 0
40  READ X
50  IF X < 0 GOTO 90
60  SUM = SUM + X
70  I = I + 1
80  GOTO 40
90  PRINT "AVERAGE SCORE IS " SUM/I
100  END
```

To verify this program by hand, first make a table of the variables that are used in the program. Now follow the actual code, line by line, writing in the values assigned to each variable. If the program cycles, write each cycle on a separate line. (It may take some thought to do this in the neatest way.) Here is the hand verification for the above program:

| SUM | I | X |
|---|---|---|
| 0 | 0 | 14 |
| 14 | 1 | 28 |
| 42 | 2 | −1 |

Prints SUM/I = 42/2 = 21

Read Program 3 line by line and check that it actually operates exactly as shown in the above verification table. When you have done this, you will be absolutely certain that Program 3 is correct, for any list of nonnegative DATA in line 10.

Obviously you can't step through a long, complex program in full detail—otherwise you wouldn't need a computer. But you can use the above method to check that loops are starting and ending properly, and that they are computing the right numbers at each step. We strongly recommend hand verification of loops—you'll be surprised how easy and reassuring it is.

In fact, good programming practice will make it easy to hand-verify your entire programs. Writing "structured programs" will also help; we give some hints below.

Suppose you have written a program that seems to work. Even though you've carefully verified it, there still might be a subtle bug. To test the program, try to find some simple cases where the outcome is predictable, and check that they do work as predicted. Also try various sensitivity tests. For example, in a behavioral model, an increase in food availability should never decrease fitness.

You've written a program, but can't get it to produce the right results. Verification has failed to detect any errors. You've checked all variable names for spelling, and formulas seem to be properly typed. Now what?

Try printing out intermediate results of the computation. In BASIC you just insert PRINT commands in appropriate places. Check that data are correctly read into variables. If you have a maximization segment, check that it *is* finding the maximum. Debugging takes imagination, and perseverance. When you finally locate the bug you'll probably wonder how you could have been so stupid; almost everyone has had this experience.

*Exercise 6:* Assume that an array $C$ of dimension $N$ contains the coefficients $c_0, c_1, \ldots, c_N$, i.e. $C(J) = c_J$. Also assume that the value of $X$ is already specified. What does the following program segment do?

```
500  Y = C(N)
510  FOR I = 1 TO N
520       Y = X * Y + C(N-I)
530  NEXT I
```

"Run" the segment by the last hand-verification method, starting with

| I | Y |
|---|---|
| - | C(N) |
| 1 | ? |
| ⋮ | ⋮ |

(This is the most efficient way to compute polynomials; an excellent place to learn good computing techniques is R.L. Burden and J.D. Faires, 1985, *Numerical Analysis*, 3rd edition, Prindle, Weber & Schmidt, Boston.)

## Flexibility

Having successfully written a program and having used it to generate the desired results, you will probably then be interested in modifying the program in some way. For example, you may wish to study the effects of changes in the parameter values (this is called a "sensitivity analysis" of your model). At some stage you may wish to modify the structure of the model itself.

Good programming habits make it easy to effect such changes in your program. For example, *all* program parameters, including the dimensions of arrays (e.g. capacity, number of patches—see

Program 6 below), should be entered in the parameter-input section (subroutine) of your program. Then in order to alter any particular parameter you will only have to type a single change in your code.

The program segment

```
NPATCH = 3
FOR I = 1 TO NPATCH
  (statements)
NEXT I
FOR J = 1 TO NPATCH
  (more statements)
NEXT J
```

is much preferable to

```
FOR I = 1 TO 3
  (statements)
NEXT I
FOR J = 1 TO 3
  (more statements)
NEXT J
```

Other features that make programs flexible include structured programming and good documentation (see below).

By the way, when making changes in a program to reflect alterations of the model, be sure to save the original bug-free code. Use the commands LOAD and SAVE to make file copies (with appropriate names) of the old and new versions of your code.

## Structured Programming

If you take a course in computer programming, your instructor will probably teach you about "structured programming." What this means, mainly, is that your big programs are broken down into simple, independent subprograms, which are coded as subroutines. Since the subroutines are short and single-purpose, they are easy to get right, and once perfected, can be used in many different programs. A really good structured program consists almost entirely of subroutines. The main program then just calls up the necessary subroutines. If you are going to do dynamic behavioral modeling regularly, we recommend you learn structured programming. There are many excellent books to help you, for example J. Blankenship, 1987, *Structured* BASIC *Programming with Technical Applications*, Prentice Hall, Englewood Cliffs, N.J.

Besides breaking your program down into relatively simple subroutines, there are various other things you can do to make programming easier and more reliable. An important step is *documentation*: intersperse informative remarks liberally in your program (but don't overdo it). Use *indentation* to identify statements within loops, and further indentation for subloops. Never type long, complicated formulas on a single line. Instead of

```
100  Y = LAMBDA * (F1(X-ALPHA+FOOD/GROUPSIZE))
         + (1-LAMBDA) * (F1(X-ALPHA))
```

try coding it as

```
100  G1 = F1(X - ALPHA + FOOD/GROUPSIZE)
110  G2 = F1(X - ALPHA)
120  Y = LAMBDA * G1 + (1-LAMBDA) * G2
```

**Example: The Patch Selection Program**

In order to illustrate the ideas we've just discussed in the context of the models of this book, we will now explicitly provide the code for the simplest patch selection problem discussed in Chapter 2. Recall that the dynamic programming equation is

$$F(x,t,T) = \max_i (1-\beta_i)[\lambda_i F(x_i', t+1, T) + (1-\lambda_i) F(x_i'', t+1, T)]$$

where $x_i' = \text{chop}(x - \alpha_i + Y_i; x_c, C)$, $x_i'' = \text{chop}(x - \alpha_i; x_c, C)$.

Program 6 is a BASIC code for this patch selection problem. The code given here is not necessarily the most efficient, so that experienced programmers may find fault with it. On the other hand, this addendum is not written for experienced programmers. But note that the program does make use of good programming habits, such as structured programming, documentation, echoing of parameters, and once-only input of parameter values.

### *Program 6*

```
100 REM PATCH SELECTION MODEL (SECTION 2.3)
110 GOSUB 1000: REM            **** ENTER PARAMETERS ****
120 DIM FO(CAPACITY), F1(CAPACITY)
130 REM FO TRACKS F(X,t,T) AND F1 TRACKS F(X,t+1,T)
```

```
140 GOSUB 2000: REM          **** INITIALIZE F1 ****
150 DIM PSTAR(CAPACITY)
160 REM PSTAR IS THE OPTIMAL PATCH
170 DIM RHS(NPATCH)
180 REM RHS(I) IS RIGHT-HAND SIDE OF DPE FOR PATCH I
190 REM START ITERATIONS
200 FOR T = HORIZON - 1 TO 1 STEP - 1
210   GOSUB 3000: REM        **** SOLVE DPE ****
220   GOSUB 4000: REM        **** PRINT RESULTS ****
230 REM UPDATE F1
240   FOR J = XCRITICAL + 1 TO CAPACITY
250     F1(J) = F0(J)
260   NEXT J
270 NEXT T
280 END: REM                 **** END OF PROGRAM ****
290 REM
1000 REM          **** INPUT & PRINT OUT PARAMETERS ****
1010 CAPACITY = 20
1020 XCRITICAL = 3
1030 ALPHA = 1
1040 HORIZON = 20
1050 NPATCH = 3: REM       NPATCH IS NO. OF PATCHES
1060 DIM BETA(NPATCH), LAMBDA(NPATCH), Y(NPATCH)
1070 DATA 0, .004, .02:REM   ****BETA VALUES****
1080 FOR I = 1 TO NPATCH
1090   READ BETA(I): NEXT I
1100 DATA 1, .4, .6:REM      ****LAMBDA VALUES****
1110 FOR I = 1 TO NPATCH
1120   READ LAMBDA(I): NEXT I
1130 DATA 0, 3, 5:REM        ****Y VALUES****
1140 FOR I = 1 TO NPATCH
1150   READ Y(I): NEXT I
1160 LPRINT "PROGRAM PATCH1"
1170 LPRINT "PARAMETER VALUES:"
1180 LPRINT "CAPACITY =" CAPACITY
1190 LPRINT "XCRITICAL =" XCRITICAL
1200 LPRINT "ALPHA =" ALPHA
1210 LPRINT " I           BETA          LAMBDA        Y"
1220 LPRINT "--------------------------------------------"
1230 FOR I = 1 TO NPATCH
1240   LPRINT I, BETA(I), LAMBDA(I), Y(I)
```

```
1250 NEXT I
1260 LPRINT "------------------------------------------"
1270 LPRINT
1280 RETURN: REM              **** END OF INPUT ****
1290 REM
2000 REM                **** INITIALIZATION SUBROUTINE ****
2010 REM OBJECTIVE IS PROBABILITY OF SURVIVAL
2020 FOR I = 0 TO XCRITICAL
2030   F1(I) = 0
2040 NEXT I
2050 FOR I = XCRITICAL + 1 TO CAPACITY
2060   F1(I) = 1
2070 NEXT I
2080 FOR I = 0 TO CAPACITY
2090   F0(I) = 0
2100 NEXT I
2110 RETURN: REM           **** END OF INITIALIZATION ****
2120 REM
3000 REM                      **** SOLVE DPE ****
3010 REM CYCLE OVER X
3020 FOR X = XCRITICAL + 1 TO CAPACITY
3030   REM COMPUTE RHS(I)
3040   FOR I = 1 TO NPATCH
3050     XPRIME = X - ALPHA + Y(I)
3060     IF XPRIME > CAPACITY THEN XPRIME = CAPACITY
3070     IF XPRIME < XCRITICAL THEN XPRIME = XCRITICAL
3080     X2 = X - ALPHA
3090     IF X2 < XCRITICAL THEN X2 = XCRITICAL
3100     TERM1 = LAMBDA(I) * F1(XPRIME)
3110     TERM2 = (1 - LAMBDA(I)) * F1(X2)
3120     RHS(I)= (1 - BETA(I)) * (TERM1 + TERM2)
3130   NEXT I
3140   VMAX = 0: REM    NOW FIND THE OPTIMAL PATCH
3150   IMAX = 0
3160   FOR  I = 1 TO NPATCH
3170     TEST = RHS(I)
3180     IF TEST <= VMAX THEN 3200
3190     VMAX = TEST: IMAX = I
3200   NEXT I
3210   F0(X) = VMAX
3220   PSTAR(X) = IMAX
```

```
3230 NEXT X
3240 RETURN: REM                    **** END OF SOLVING DPE ****
3250 REM
4000 REM                            **** PRINT RESULTS ****
4010 LPRINT "TIME:" T
4020 LPRINT
4030 LPRINT "  X             F(X,t,T)    Opt Patch"
4040 LPRINT "----------------------------------"
4050 FOR X = XCRITICAL TO CAPACITY
4060   LPRINT X, F1(X), PSTAR(X)
4070 NEXT X
4080 LPRINT "----------------------------------"
4090 RETURN: REM        **** END OF PRINTING RESULTS ****
```

Program 6 follows the pseudocode of Section 2.3 closely. The program itself occurs on lines 100-280. It is a structured program, in the sense that most of the actual computations are placed in subroutines. For example, line 110 calls the parameter input subroutine starting at line 1000. This subroutine also echoes the parameters to the printer, so your output will include a list of parameters; this is always useful for later reference (and for checking that the computer has read the correct parameter values). A simple modification of this subroutine can be used to make any desired changes to the parameter values.

Note also the helpful documentation provided by the remarks (some in asterisks). Also note that remarks can be placed on the same line as program statements, by using a colon (:). Colons are used in BASIC to include several program statements on a single line—see line 1090, for example.

Line 120 dimensionalizes the "new" and "old" fitness functions $F0$ and $F1$, and line 130 reminds you which is which. The terminal fitness function $\phi(x)$ given by

$$\phi(x) = \begin{cases} 0 & \text{for } 0 \le x \le x_c \\ 1 & \text{for } x_c < x \le C \end{cases}$$

is entered via the initialization subroutine starting at line 2000 (and called on line 140). This subroutine puts `F1(X)=`$\phi$`(X)`, and also initializes `F0(X)=0`. Most BASIC systems automatically initialize array values to zero, but it is a good idea to make sure (remember that we need to have $F0(x_c) = 0$ always).

Line 150 dimensionalizes `PSTAR(X)`, which will be an array which records the optimal patch number $I^*(X)$ at each iteration. Line

170 dimensionalizes `RHS(I)`, which will be used to store the expression

$$\text{RHS}(i) = (1 - \beta_i)[\lambda_i F(x_i', t+1, T) + (1 - \lambda_i) F(x_i'', t+1, T)]$$

The actual iterations of the dynamic programming equation begin on line 200. Two subroutines are called, on lines 210 and 220. The first subroutine solves the dynamic programming equation (see below), and the second one simply prints the results. Finally lines 240-260 update `F1(X)` in preparation for the next iteration, and the program is finished.

Before describing the subroutines 3000 and 4000, let us pause to hand-verify that our program works correctly, assuming for the moment that all the subroutines have been correctly coded. (We write `T = HORIZON` for simplicity.)

| $N$ | `F0(X)` | `F1(X)` |
|---|---|---|
| 0 | 0 | $\phi(X) = F(X,T,T)$ by initialization |
| 1 | $F(X,T-1,T)$ by solving DPE | $F(X,T-1,T)$ by lines 240-260 |
| 2 | $F(X,T-2,T)$ by solving DPE | $F(X,T-2,T)$ by lines 240-260 |
| ⋮ | | |
| $T-1$ | $F(X,1,T)$ by solving DPE | |

It looks as if the program should work correctly.

Next let us examine the subroutine starting on line 3000. This is the core of the program. For each $X$ from $x_c + 1$ to $C$ the subroutine calculates

$$\max_{1 \le i \le N_{\text{patch}}} \text{RHS}(i) = F(X,t,T) = F0(X)$$

(given that $F1(X)$ equals $F(X,t+1,T)$), which is our dynamic programming equation. Thus, lines 3030-3130 calculate `RHS(1)`, ...,`RHS(NPATCH)`—you can hand-verify that this code segment is correct. Then lines 3140-3200 find the maximum `VMAX` of these values, and keep track of the index IMAX which gives the maximum. This index specifies the optimal patch, and is one of our desired outputs. Finally on lines 3210-3220 we store the computed maximum VMAX as FO(X), and the index IMAX as PSTAR(X).

These computed values are printed out by the subroutine starting on line 4000. The instructions on lines 4020-4040, and 4080 make the output look neat.

We hope you will agree that a structured program is especially easy to read. Structured programs are also easy to modify, and easy to code in the first place, because the program has been broken down into relatively simple separate and independent subroutines.

*Exercise 7:* Note that Program 6 assumes that $\alpha_i = \alpha$ is the same for all patches. List all program line numbers where changes would be required if in fact the $\alpha_i$ varied between patches. How should Program 6 have been written so that these changes would occur in a single line?

*Exercise 8:* Suppose you want to perform 100 iterations, but not print out all the results. Modify the output subroutine in Program 6 so that

(a) only every 10th iteration is printed, or

(b) only the last 10 iterations are printed.

Hint for (a). One way to do this is to use the BASIC operator MOD, which gives the remainder upon dividing integers:

N MOD M equals the remainder upon dividing N by M

Thus the expression

```
N MOD 10 = 0
```

will be true if and only if N is a multiple of 10.

# II Applications

To go on investigating without the guidance of theories is like attempting to walk in a thick mist without a track and without a compass. We should get somewhere under these circumstances, but chance alone would determine whether we should reach a stony desert of unintelligible facts or a system of roads leading in some useful direction; and in most cases chance would decide against us.

*August Friedrich Leopold Weismann*

*Each of the next five chapters concerns a specific application of the dynamic state variable methodology that was developed in Chapter 2. Any of Chapters 3–7 can be read independently. Each of these chapters covers a field of behavioral ecology on which an entire book could be written (and many have, in fact), so that the individual reader who is an expert in the particular subject may find much of his or her subject missing in the chapter. Although the chapters are not comprehensive, each one deals with one or more behavioral problems for which the state variable approach leads to new insights and interpretations of data and suggests new experiments.*

*When reading these chapters, one must also remember that certain details of the biology will always be excluded from particular models. The objective of these models is not to reproduce all the complexities of nature within the computer; rather, they attempt to provide insight into behavioral phenomena, and to suggest new observations and experiments. Existing data are used to suggest the form of the model and roughly to estimate parameters in the model, but we do not try to "fit" the model to the data in extremely precise ways. In this kind of modeling, qualitative trends are more often important than quantitative agreement. In our opinion, the ideal situation is one in which the modeler and experimental scientist work in tandem, with experiment followed by model followed by experiment, and so on. In order to achieve such a sequence, one must use theory to predict experimental results—even if the pre-*

*dictions are at first only qualitative. Such coupling of experiment and theory is one of the fundamental bases of true advancement in science.*

# 3 The Hunting Behavior of Lions

The behavior of the African lion (*Panther leo*) has been studied intensively over the past two decades (Schaller 1972, Bertram 1975, Owens and Owens 1984, Packer 1986). The lion is the only member of the cat family that normally lives and hunts in groups (Packer 1986). In the Serengeti, a typical lion pride consists of 2–18 related adult females, their young offspring, and 1–7 unrelated adult males. A lion pride is thus a relatively stable social group.

Hunting is carried out primarily by female members of the pride, either hunting alone or in groups consisting of two to eight lions, depending in part on the particular prey species. In the Serengeti important prey species include, in approximate order of importance (Schaller 1972, Table 43): zebra (*Equus burchelli*), wildebeest (*Connochaetes taurinus*), buffalo (*Syncerus cafer*), and Thomson's gazelle (*Gazella thomsoni*). Warthogs (*Phacochoerus aethiopicus*), giraffes (*Giraffa camelopardalis*), ostriches (*Struthio camelus*), and other species are also sometimes taken.

The functional explanation usually suggested for lion sociality is the need to cooperate in overcoming large prey animals, but as Packer (1986) points out, other felines, such as cougar (*Felis concolor*) and leopard (*Panthera pardus*), also exploit large prey, but have not evolved social behavior (for information on leopard ecology see Bertram 1982). Consequently, other advantages of group hunting behavior need to be considered. In this chapter we analyze several alternative explanations of group hunting. First we summarize Schaller's data on the group hunting behavior of Serengeti lions. We then review previous explanations. Finally we develop a dynamic model of group hunting behavior, which we then apply to the Serengeti data. A modification of this model allows us to assess the benefits of conspecific scavenging, or communal sharing of kills between pride members.

Table 3.1
Success rates for lion hunts*

| group size | T. Gazelle no. hunts | T. Gazelle % success | Wildebeest & Zebra no. hunts | Wildebeest & Zebra % success | Other no. hunts | Other % success |
|---|---|---|---|---|---|---|
| 1 | 185 | 15 | 33 | 15 | 31 | 19 |
| 2 | 78 | 31 | 17 | 35 | 11 | 9 |
| 3 | 42 | 33 | 16 | 13 | 5 | 20 |
| 4-5 | 42 | 31 | 16 | 37 | 4 | 25 |
| 6+ | 15 | 33 | 21 | 43 | 7 | 0 |

*From Schaller 1972, Table 59.

## 3.1 The Serengeti Lion

Schaller (1972) observed a total of 523 chases of prey by Serengeti lions, and recorded the success rates for hunting groups of different sizes—see Table 3.1. Slightly more than half the hunts (274 out of 523) were undertaken by groups of lions; for zebra or wildebeest prey, nearly 70% of the hunts were by groups. With some exceptions, groups experienced a higher success rate than individuals. If, as a first approximation, we ignore the fact that females must usually share their kills with males and young, then the average individual food recovery per chase $\bar{f}$ can be expressed as

$$\bar{f} = \frac{p_n \overline{W}}{n} \tag{3.1}$$

where $n$ = hunting group size, $p_n$ = probability of successful hunt, and $\overline{W}$ = average weight of prey. Note from Table 3.1 that this measure of success gives a slight advantage to groups of size two (relative to single hunters) for Thomson's gazelle, and a somewhat larger advantage for wildebeest and zebra. Groups larger than two are never advantageous on these terms, yet larger groups are frequently observed (Table 3.2).

A more detailed analysis of average individual daily feeding rates as a function of hunting group size was given by Caraco and Wolf (1975), who included such additional features as multiple kills, scavenging by hyenas, and ecological circumstances

Table 3.2
Average number of adult lions feeding on carcasses*

| Area | Wildebeest | Zebra | T. gazelle |
|---|---|---|---|
| plains | 3.7 | 3.7 | - |
| seronera | 6.9 | 7.3 | 2.0 |
| woodlands | 4.7 | 4.0 | 1.7 |

*From Schaller 1972, Table 40.

(habitat and season). In all cases it turned out that the optimal group size in terms of individual feeding rates was two lions per group. Packer (1986) has reconsidered Schaller's data, and criticizes Caraco and Wolf's assumption that multiple kills are as likely for groups of two hunters as for larger groups. Packer also discusses observations of the stomach contents of lions belonging to prides of various sizes, and finds no increase for larger prides (Packer 1986, Figure 4). He concludes that "... there are no good data showing that cooperative hunting is in fact beneficial to lions ... A solitary lion achieves rates of food intake at least as high as a group member."

## 3.2 Some Possible Explanations of Lions' Hunting Behavior

Given this information, how can one explain the social hunting behavior of lions? Two things are in need of explanation—the evolution of social rather than solitary hunting, and the fact that hunting groups are often larger than the apparent optimum group. Regarding the latter point, note that the group sizes listed in Table 3.2 refer to groups of lions observed feeding at carcasses. Since male and juvenile lions feed on carcasses killed by females, these average group sizes are larger than the average hunting group sizes. Schaller (1972, Table 2) points out that the 14 prides that he observed contained a total of 82 adult females plus 30 adult males and 10 subadults, together with a large number of cubs. In the computations described later we use the ratio of $q = 122/82 \approx 1.5$ feeders to hunters for large carcasses, although we recognize that in some cases the ratio may be much greater than this. The pos-

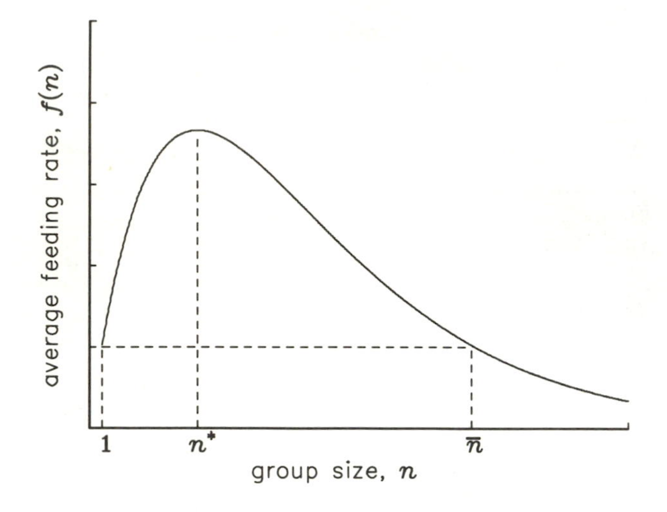


Figure 3.1 Stable vs. optimal group size.

sibility that hunters may also scavenge the kills of other hunting groups from the same pride is discussed in Section 3.4.

A possible explanation for group hunting behavior is that individual feeding rate is not the only factor involved in group hunting, and some other measure such as breeding success ought to be considered (Caraco and Wolf 1975). But since hunting groups are temporary subgroups of a given pride, breeding success is likely to be influenced by pride size, not hunting group size. Similarly, prides may be more successful at defending territories than individuals or mated pairs, but again this does not seem to explain group hunting.

Suppose for the moment that, as suggested by Caraco and Wolf (1975), the individual feeding rate $\bar{f} = \bar{f}(n)$ is peaked at $n = n^* > 1$ (Figure 3.1). Assume a group of $n^*$ lions begins hunting. An additional lion then faces the choice of hunting alone or joining the existing group. If $\bar{f}(n^* + 1) > \bar{f}(1)$, as is the case in Figure 3.1, then the extra individual will do better by joining the group than by hunting alone. The stable group size $\bar{n}$, in the sense that further additions will not be attracted, satisfies $\bar{f}(\bar{n}) = \bar{f}(1)$.* The stable group $\bar{n}$ may be much larger than the optimal group $n^*$

* More accurately $\bar{n}$ is determined by the conditions $\bar{f}(\bar{n}) \geq \bar{f}(1)$, $\bar{f}(\bar{n} + 1) < \bar{f}(1)$.

(Sibly 1983, Pulliam and Caraco 1984, Slobodchikoff 1984, Clark and Mangel 1984). Giraldeau (1986) reconsiders Schaller's data, making an additional adjustment to $\bar{f}(n)$ to account for relatedness of pride members and inclusive fitness, and calculates that the stable hunting group size is three adult females for all cases except gazelle, where it is two. However, over a third of the groups observed hunting zebra had more than three members (Table 3.1), so that some discrepancy between theory and data remains. There is also the problem of explaining why nearly half the observations were of lions feeding alone.

Another suggested advantage of large hunting groups is based on considerations of the variance of daily food intake (Caraco 1981). A solitary lion hunting zebra, for example, would average only one kill every 6.7 days. Female lions need about 5 kg of meat per day, and have stomach capacity of about 22 kg (Schaller 1972, p. 276). Thus without relying on fat reserves, a female lion cannot survive more than 4–5 days without food. Group hunting increases the probability of a successful kill (two lions hunting zebra average one kill about every 3 days, and six lions average one kill every 2.3 days, according to Table 3.1). The question then becomes one of evaluating the tradeoff between mean and variance in daily food intake.

A very simple model of the mean-variance tradeoff can be constructed as follows (Pulliam and Millikan 1982, Clark and Mangel 1986). Assume that an individual forager captures food items according to a Poisson process with parameter $\lambda$.* If handling time is negligible, then the probability of encountering $k$ items of food in time $t$ is given by

$$p(k,\lambda) = \frac{(\lambda t)^k}{k!}e^{-\lambda t} \qquad k = 0, 1, 2, \ldots \tag{3.2}$$

The mean and variance of this distribution are both equal to $\lambda t$.

Now suppose that a group of $n$ foragers encounters food items at the combined rate $n\lambda$. Then the probability that the group encounters $k$ items in time $t$ is equal to $p(k, n\lambda)$. Assuming that food items are shared equally, the individual's mean food consumption is again $\lambda t$, but the variance is $\lambda t/n$.

Finally, suppose that the forager requires $R$ items of food in order to survive a certain period. Then (assuming equal division

* See Appendix 1 to Chapter 1 for the mathematics underlying the Poisson process.

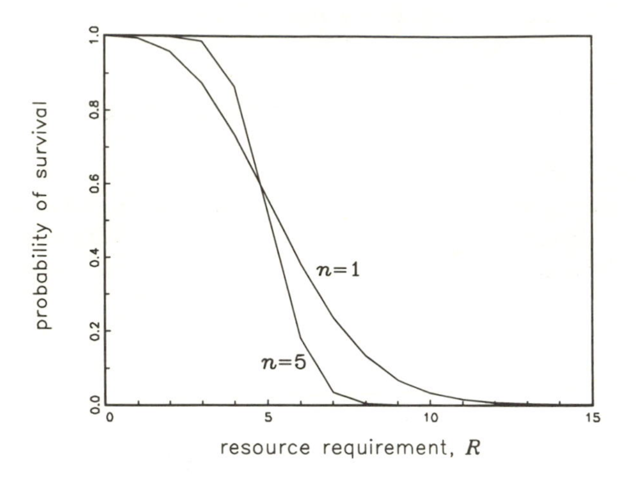


Figure 3.2 Probability of survival for Poisson search model ($\lambda t = 5.0$), as a function of resource requirement $R$ (from Clark and Mangel 1986).

of food) the individual will survive provided that the group encounters $R$ food items:

$$\text{Pr (survive period)} = \Pr(k \geq nR) = \sum_{k=nR}^{\infty} p(k, n\lambda)$$

This function (of $R$) is plotted in Figure 3.2, for $\lambda t = 5.0$ and for $n = 1$ and $n = 5$. Note that the two curves cross over at $\mu = \lambda t = 5$; for $R < \mu$ the larger group is preferable, while for $R > \mu$ the smaller group does better.

We can describe this result by saying that the forager should be risk-averse (i.e. should favor low variance) whenever expected food intake exceeds requirements, and should be risk-prone when expected food intake falls short of requirements. Such risk sensitivity ("when desperate, gamble") is in fact a general principle of strategic behavior under uncertainty (Dubins and Savage 1976). Its implications for behavioral ecology have been investigated by Caraco (1981), Stephens and Charnov (1982), McNamara and Houston (1986), Real and Caraco (1986), and others.

Members of large foraging groups generally experience lower variance in food intake than lone foragers (Caraco 1981). The qualitative prediction of this analysis is therefore that among species that forage socially, foraging group size will be an increasing function of resource abundance. Similarly, social foraging behavior should break down when food becomes scarce. The observations of Kalahari lions by Owens and Owens (1984) give support for this prediction.

The reader will perceive that the above model cannot readily be used to generate quantitative predictions of foraging group size. For example, whether a lion survives for, say, six months depends on the sequence of daily food intakes over the entire period. One cannot simply specify a minimum food requirement $R$ for a long period of time. What is required is a model that allows foraging decisions to respond dynamically to the lion's current degree of hunger and to current food availability.

## 3.3 A Dynamic Model

The following dynamic model of lion hunting appears in Clark (1987). The state variable $X(t)$ represents gut contents (kg of meat) of an individual adult female lion at the end of day $t$. Gut contents are constrained by capacity $C$:

$$0 \leq X(t) \leq C. \tag{3.3}$$

The value $X(t) = 0$ is assumed to correspond to death by starvation. The strategies to be considered are simply the size $n$ of hunting group that the lion participates in on day $t$, with $n = 1, 2, \ldots, N_{\max}$.

Let $\alpha_n$ denote the daily metabolic cost for a lion hunting in a group of size $n$. Let $Z$ denote food captured per individual lion per day. For a hunting group of size $n$ let

$$\Pr(Z = z_{nj}) = \lambda_{nj} \qquad j = 1, 2, \ldots, J. \tag{3.4}$$

For example, if only one chase is attempted per day, $p_n$ is the probability of success, and $W$ is the edible weight of one prey, then we have

$$\left.\begin{aligned} \Pr(Z = \frac{W}{qn}) &= p_n \\ \Pr(Z = 0) &= 1 - p_n \end{aligned}\right\} \tag{3.5}$$

where $q$ is the ratio of feeders to hunters and the prey is assumed to be shared equally between feeders. If food size $W$ is a random variable with, say

$$\Pr(W = w_j) = \lambda_j \qquad j = 1, 2, \ldots, J$$

then (3.4) becomes

$$\left.\begin{aligned} \Pr(Z = \frac{w_j}{qn}) &= p_n\lambda_j \\ \Pr(Z = 0) &= 1 - p_n \end{aligned}\right\}. \tag{3.6}$$

Other details, such as multiple chases per day, or multiple kills per chase, can also be included in the model by adjusting the probability distributions $\lambda_{nj}$ accordingly, as will be seen below.

Our model will pertain to hunting behavior during the non-breeding season of $T$ days; we will assume a terminal fitness function $\phi(X(T))$. As in Chapter 2, the lifetime fitness function is

$$F(x, t, T) = \max E\{\phi(X(T)) \mid X(t) = x\} \tag{3.7}$$

where the maximum is taken over $n$ = hunting group size. The dynamic programming equation is

$$F(x, t, T) = \max_n \Sigma_j \lambda_{nj} F(x'_{nj}, t + 1, T) \tag{3.8}$$

$$x'_{nj} = \text{chop}(x - \alpha_n + z_{nj}; 0, C). \tag{3.9}$$

The end condition is

$$F(x, T, T) = \phi(x). \tag{3.10}$$

In order to solve the dynamic programming equation we need values for the model parameters. We discuss two examples, for prey consisting of Thomson's gazelle or zebra respectively.

### Thomson's Gazelle

The following parameter values are taken from Schaller's (1972) estimates:

daily food requirement: $\alpha = 6\,\text{kg}$

gut capacity: $C = 30\,\text{kg}$ (average for males and females)

mean edible biomass per gazelle: $W = 11.25\,\text{kg}$

Table 3.3
Average daily individual food intake $Z_n$ for Serengeti lions hunting Thomson's gazelle

| Group Size $n$ | Average Food Intake/Day |
|---|---|
| 1 | 5.4 kg |
| 2 | 5.8 kg |
| 3 | 3.8 kg |
| 4 | 2.9 kg |

Lions make up to three chases on gazelle per day. If $p_n$ is the kill probability of a successful chase then the number of kills per day is a binomial random variable (see Appendix 1.3):

$$\Pr\left(Z = \frac{kW}{n}\right) = \lambda_{kn} = \binom{3}{k} p_n^k q_n^{3-k} \qquad k = 0, 1, 2, 3 \tag{3.11}$$

where

$$q_n = 1 - p_n.$$

From Table 3.1 we have

$$p_1 = 0.15,\ p_2 = 0.31,\ p_n = 0.33 \text{ for } n \geq 3. \tag{3.12}$$

(The success rate in hunting gazelle does not appear to increase for $n > 3$.)

The average daily individual food intake, as a function of group size $n$ is shown in Table 3.3. As noted previously, in order to maximize average daily individual food intake, lions should always hunt in groups of size $n = 2$. On the other hand, the optimal hunting group sizes $n^*$ as computed from our dynamic model (i.e. by solving Eq. (3.8) on the computer*) are shown in Table 3.4.

* A technical point arises in the computer solution of Eq. (3.8), since the state variable $x$ now takes on fractional rather than integer values. We used simple linear interpolation to compute the values of $F(x'_{nj}, t+1, T)$ for noninteger $x'_{nj}$ from the stored array $F(x, t+1, T)$ with $x = 0, 1, \ldots, C$. This procedure is described in Section 8.3.

Table 3.4
Thirty-day survival probabilities $F(x, 1, 30)$ and optimal foraging group sizes $n^*$ on day 1, for Serengeti lions hunting Thomson's gazelle

| Initial Stomach Contents $x$ kg | Optimal Group Size $n^*$ | Probability of Survival $F(x, 1, 30)$ |
|---|---|---|
| 5 | 1 | 0.06 |
| 10 | 2 | 0.13 |
| 15 | 2 | 0.19 |
| 20 | 2 | 0.29 |
| 25 | 2 | 0.31 |
| 30 | 2 | 0.35 |

We use $T = 30$ days, and $\phi(x)$ is taken as the terminal fitness function corresponding to the probability of survival, i.e. $\phi(x) = 0$ for $x = 0$, $\phi(x) = 1$ for $x > 0$. The data in Table 3.4 show the optimal group sizes on day one of thirty days. This is in fact a stationary strategy, valid for times $t$ more than seven days prior to the horizon $T$.

Note that the survival probabilities in Table 3.4 are quite low, particularly for lions that begin with low reserves $x$. According to Table 3.3 gazelle do not provide quite enough food to meet the daily 6 kg requirement. Our model predicts that few lions will survive periods during which gazelle is the only available prey. In fact, gazelle is the most common prey species for Serengeti lions from May–October, when the migratory herds of wildebeest and zebra are not present (Schaller 1972). However, various other prey species are taken at this time, and our single-species model doubtlessly underestimates the probability of survival. A more complex model of lion physiology, allowing for the use of fat reserves during periods of resource scarcity (see Appendix 2.1.4), would also likely give higher predicted survival rates.

The model also predicts that lions should adopt the risk-prone strategy of foraging alone whenever their stomach contents are below 9 kg. Under starvation conditions in which gazelle are the only prey, most lions will in fact have $x < 9$ kg, so that the majority of hunts will be by single lions. A correlation between stomach contents and hunting group size in the Serengeti seems to be un-

Table 3.5
Probabilities of survival ($T = 30$) for the dynamic optimization model, compared with the risk-sensitive model ($n = 1$), and the average food maximization model ($n = 2$). Prey is Thomson's gazelle.

| Gut Contents | Probability of Survival | | |
|---|---|---|---|
| $x$ kg | Optimum | $n = 1$ | $n = 2$ |
| 5 | 0.063 | 0.022 | 0.058 |
| 10 | 0.129 | 0.044 | 0.119 |
| 15 | 0.195 | 0.066 | 0.183 |
| 25 | 0.314 | 0.099 | 0.305 |
| 30 | 0.354 | 0.100 | 0.346 |

known, although Owens and Owens (1984) report that lion prides in the Kalahari break up into individual nomadic hunters during the dry season, when prey becomes scarce.

The survival probabilities resulting from suboptimal hunting strategies with fixed group size $n = 1$ or 2, respectively, are shown in Table 3.5. The risk-prone strategy ($n = 1$) is strongly suboptimal, but the average food intake maximizing strategy ($n = 2$) is only marginally suboptimal.

None of the models discussed predicts that lions will sometimes hunt gazelle in groups larger than two, although the data in Table 3.1 indicate that in fact 22% of the gazelle hunts observed by Schaller were made by larger groups. There are various reasons why this might occur. Lions observed chasing gazelle may have originally expected other prey species. Group hunts can sometimes result in multiple kills: Schaller (1972, p. 254) reports that 18 successful gazelle hunts by groups of four or more lions resulted in four multiple kills of 2 or 3 gazelle. By making some explicit assumptions about the effect of group size on the probability of multiple kills, one could modify the above model, in attempting to understand why larger hunting groups sometimes form while hunting gazelle.

## Zebra

Zebra are much larger animals than Thomson's gazelle, with an average edible biomass of 164 kg, which is large enough to meet

Table 3.6
Thirty-day survival probabilities $F(x, 1, 30)$ and optimal group sizes $n^*$ on day one, for Serengeti lions hunting zebra: (a) one chase per day; (b) up to three chases per day.

| Initial Stomach Contents, $x$ kg | Optimal Group Size, $n^*$ | | Probability of Survival, $F(x, 1, 30)$ | |
|---|---|---|---|---|
| | (a) | (b) | (a) | (b) |
| 5 | 4 | 6 | 0.171 | 0.806 |
| 10 | 5 | 6 | 0.281 | 0.957 |
| 15 | 6 | 6 | 0.347 | 0.989 |
| 20 | 6 | 6 | 0.387 | 0.994 |
| 25 | 6 | 6 | 0.412 | 0.995 |
| 30 | 6 | 6 | 0.415 | 0.995 |

the daily metabolic requirements of 27 lions! This consideration alone makes one suspicious of the "optimality" of hunting zebra in groups of two lions.

To begin, let us use the same model as for the case of gazelle hunts, with the following changes in parameter values:

$$W = 164\,\text{kg}$$
$$p_1 = 0.15, \quad p_2 = 0.33, \quad p_3 = 0.37,$$
$$p_4 = 0.40, \quad p_5 = 0.42, \quad p_6 = 0.43$$

These probabilities are obtained by smoothing the data given in Table 3.1; we assume that $p_n = 0.43$ for $n \geq 6$. The dynamic optimization results ($T = 30\,\text{days}$) are shown in Table 3.6.

Note that the optimal group sizes are now much larger than $n = 2$. The explanation should be clear by now: two lions cannot possibly consume a whole zebra in one day. In fact, the maximum possible daily intake, for a very hungry lion ($x = 0$) is $C + \alpha = 36\,\text{kg}$. Hence one zebra will always feed at least four lions for one day.

This analysis immediately raises further questions. Doesn't a zebra carcass last for more than one day? Can't lions undertake more than one chase of zebra per day? We will now describe several modifications of the basic model designed to address these effects.

First, if unsuccessful in the first chase, lions may attempt a second or third chase in the same day; once a kill is made, the group spends the remainder of the day feeding. If $p_n$ is the probability of success in a single chase, and if up to three chases may be made per day, then the probability of success per day becomes

$$p'_n = p_n + q_n p_n + q_n^2 p_n \tag{3.13}$$

(because the probability of success on the first chase is $p_n$, while the probability of being unsuccessful on the first chase and successful on the second chase is $q_n p_n$, and so on). The results of this modification are shown in the (b) columns of Table 3.6. Survival probabilities are significantly increased relative to the case of single chases per day, and the optimal group size is now always $n = 6$ (groups of size 6 have an 81% probability of success per day, if multiple chases are possible, and this means that risky hunting in small groups is no longer optimal).

Suppose next that six well-fed lions ($x = C$) kill a zebra, and 9 lions feed on the carcass. Then 54 kg of meat are consumed, leaving 110 kg uneaten. Zebra carcasses can last up to three days (Schaller 1972), so that it might be worthwhile for the lions to protect or hide the carcass from scavengers, although this is apparently seldom done in the Serengeti (S. Cairns, personal communication). Our model can easily be adapted to include this possibility, by incorporating a delay effect for consuming zebra. One needs to formulate a specific rule for modeling the utilization of "leftovers" from a partially eaten carcass.

For example, let us suppose that a group of $n$ hunters ($1.5n$ feeders) will feed on a leftover zebra carcass rather than initiate a new hunt, if and only if there is enough meat remaining to feed the entire group to capacity (and provided that the carcass is less than four days old). Let $x'_n$ denote stomach contents when the carcass is abandoned, and let $\tau_n$ be the number of days spent feeding on the carcass. Expressions for $x'_n$ and $\tau_n$ can be derived as follows. First let $f_n = W/(1.5n)$, i.e. $f_n$ is the food available per lion from a carcass of weight $W$. On the first day each lion can eat $C + \alpha - x$. Hence if $f_n < C + \alpha - x$ the group eats the whole carcass, and we have

$$\left.\begin{aligned} x'_n &= x - \alpha + f_n \\ \tau_n &= 1 \end{aligned}\right\} \qquad \text{if } f_n < C + \alpha - x.$$

Next, if $f_n \geq C + \alpha - x$ then the lions can feed to capacity. If $f_n < C + 2\alpha - x$ there is not enough food for two days, so that

Table 3.7
Effects of utilizing carcasses for more than one day. Thirty-day survival probabilities $F(x,1,30)$ and optimal hunting group sizes $n^*$ on day one for Serengeti lions hunting zebra: (a) one chase per day; (b) up to three chases per day.

| Initial Stomach Contents, $x$ kg | Optimal Group Size, $n^*$ | | Probability of Survival, $F(x,1,30)$ | |
|---|---|---|---|---|
| | (a) | (b) | (a) | (b) |
| 5 | 4 | 6 | 0.234 | 0.808 |
| 10 | 5 | 6 | 0.382 | 0.962 |
| 15 | 4 | 6 | 0.471 | 0.991 |
| 20 | 3 | 6 | 0.529 | 0.997 |
| 25 | 4 | 6 | 0.570 | 0.998 |
| 30 | 6 | 6 | 0.579 | 0.998 |

by assumption the carcass is abandoned after the first day:

$$\left.\begin{aligned} x'_n &= C \\ \tau_n &= 1 \end{aligned}\right\} \qquad \text{if } C + \alpha - x \leq f_n < C + 2\alpha - x.$$

Similarly

$$\left.\begin{aligned} x'_n &= C \\ \tau_n &= 2 \end{aligned}\right\} \qquad \text{if } C + 2\alpha - x \leq f_n < C + 3\alpha - x$$

and

$$\left.\begin{aligned} x'_n &= C \\ \tau_n &= 3 \end{aligned}\right\} \qquad \text{if } C + 3\alpha - x \leq f_n$$

(recall that the carcass is abandoned after three days). As usual, let $F(x,t,T)$ denote expected lifetime fitness for a lion having $X(t) = x$, and not possessing any leftover carcass on day $t$. The dynamic programming equation now becomes

$$\begin{aligned} F(x,t,T) = \max_n \{ & p_n F(x'_n, t+\tau_n, T) \\ & + (1-p_n) F(x-\alpha, t+1, T)\} \end{aligned} \tag{3.14}$$

(with $t + \tau_n$ replaced by $T$ if $t + \tau_n > T$). The results of this modification are shown in Table 3.7.

Comparing Tables 3.6 and 3.7, we see that the use of carcasses over a three-day period has a considerable effect if lions only undertake a single zebra chase per day (optimal group sizes decrease, and survival probabilities increase, when leftover zebra carcasses are utilized). However, for the case of multiple chases per day, the effect of utilizing leftover carcasses is almost negligible. This prediction seems to fit with the observation that lions seldom protect carcasses for future use, but also suggests that leftover carcasses might become important when prey is scarce, so that multiple chases are less likely. B. Bertram (personal communication) has observed lions feeding on the same carcass for extended periods under these conditions.

What insights have we obtained so far from the dynamic modeling of lion behavior? First, the importance of considering lions' stomach capacity has become obvious. For large prey such as zebra, the optimal hunting group size is largely determined by stomach capacity. None of the earlier analyses of "optimal" hunting group size considered stomach capacity as a limiting constraint. Similarly, the perishability of carcasses under tropical conditions was not considered. (If prey carcasses last indefinitely, then the optimal group size predicted by the dynamic model is the group size that maximizes $\overline{f}_n$—i.e. $n = 2$.)

It is also very simple and straightforward to introduce additional behavioral complexities, such as multiple chases, use of leftovers carcasses, etc. into the basic dynamic model. The relative importance of such behavioral traits can thus easily be assessed. In a sense, our dynamic model has in fact worked too well—the predicted optimal hunting groups of six lions are considerably larger than the mean observations reported by Schaller (1972) and Packer (1986). In the next section we will discuss a further aspect of lion hunting behavior which has an important influence on the optimal hunting group size.

## 3.4 Communal Sharing

In his analysis of the evolution of lion sociality, Packer (1986) concludes that "much of lion sociality results from the fact that lions so commonly scavenge from conspecifics." The modeling framework used in this chapter provides a convenient method of quantitatively assessing the contribution that conspecific scavenging (among members of the same pride) could make towards increasing the fitness of individual lions. First let us try to understand this question intuitively.

In the previous section we saw that the combination of large prey biomass, finite predator capacity, prey perishability, and increased kill probabilities from group hunting, imply that the fitness of individual lions is maximized by cooperative hunting in groups. In the case of large prey such as zebra, the average amount of food killed per lion per day, by optimally sized groups, is less than the maximum possible, which would be obtained by hunting in groups of size two. These small groups are not capable of consuming an entire zebra, however, so that much of the zebra carcass would be wasted.

Under a communal sharing arrangement, however, the kills made by any group of hunters would be shared among the whole pride. If the pride were fairly large, the stomach capacity limitation would no longer be pertinent. The intuitive result then is that, under these conditions, the optimal hunting group size would be two (or whatever group size results in the maximum rate of food killed). Packer (1986) therefore hypothesizes that scavenging the kills of pridemates was sufficiently advantageous to engender the evolution of social behavior in lions.

We now discuss modifications of the group-hunting model required to investigate this hypothesis. Let $N$ denote the size of the pride (number of adult females), and let $n$ denote the size of hunting groups. For simplicity we only consider cases in which $n$ divides $N$; let $m = N/n$ denote the number of hunting groups. Let $p_n$ denote the daily kill probability per hunting group. If $Z$ denotes total daily kill by the $m$ hunting groups of the pride, we have

$$\Pr(Z = jW) = \binom{m}{j} p_n^j q_n^{m-j} \qquad j = 0, 1, \ldots, m \tag{3.15}$$

Now assume that prides always hunt in groups that maximize individual fitness (probability of survival), given that kills are shared equally among all pride members. The dynamic programming equation for individual fitness $F(x, t, T)$ now becomes

$$F(x, t, T) = \max_n \sum_{j=0}^{m} \binom{m}{j} p_n^j q_n^{m-j} F(x'_j, t+1, T) \tag{3.16}$$

Table 3.8
Maximum 30-day survival probabilities, $F(x, 1, 30)$ and optimal hunting group sizes $n^*$, for prides of Serengeti lions hunting zebra employing communal sharing.

| Pride Size, Females | Probability of Survival $F(30, 1, 30)$ | Optimal Hunting Group Size $n^*$ |
|---|---|---|
| 2 | 0.18 | 2 |
| 4 | 0.75 | 2 |
| 6 | 0.95 | 2 |
| 8 | 0.993 | 2 |
| 10 | 0.999 | 2 |

where

$$x'_j = \text{chop}(x - \alpha + \frac{jW}{qN}; 0, C). \tag{3.17}$$

In Eq. (3.16) the values of $n$ divide $N$, and $m = N/n$.

Table 3.8 shows the 30-day survival probabilities and optimal hunting group sizes with zebra as prey (one chase per day), for lion prides consisting of from 2–12 adult females. As one would expect, under communal sharing the optimal hunting group size is always $n = 2$ (the average-rate-maximizing group size). The table indicates that communal sharing has a marked influence on individual survival rates. For example, if six lions hunt in a single group their survival probability is 42% (Table 3.6) but if they hunt in groups of two and share kills, this rises to 95%. Thus the same conditions that favor group foraging also strongly favor the evolution of communal sharing.

It is important to stress that this argument does not involve group selection; only individual fitness is assumed in the calculation. Packer (1986) notes that when a lion hunting group kills a large prey, pride members from other hunting groups often can detect this fact, and converge to scavenge the kill. Presumably the open nature of the Serengeti savannah facilitates this sharing behavior.

Some degree of social cooperation is probably needed, however, to prevent hunting groups from exceeding the optimum size. The fact that, for large prey, the hunting groups observed by Schaller somewhat exceeded two lions per group (see Giraldeau 1986) may indicate either that social cooperation is imperfect in lion prides, or that there are additional benefits associated with larger hunting groups.

Table 3.8 may suggest that survival probabilities would continue to increase with pride size. However, it is clear that prides with more members than can be nourished by a typical large carcass would not be optimal. An average zebra (164 kg), for example, would provide the daily meat requirements of at most 27 lions. Because of interference and communication problems with a large number of hunting groups operating communally, it seems likely that the optimal pride size is considerably less than this limit. Packer et al. (1987) observed that, in the Ngorongoro crater, per capita reproductive success was higher for prides containing 3–10 adult females than for either smaller or larger prides. The 14 Serengeti prides studied by Schaller (1972) contained an average of 5.9 adult females per pride.

## 3.5 Discussion

No single model can hope to capture all the complexity of a real-world behavioral situation. The dynamic, state variable modeling approach provides at least two important advantages over other modeling frameworks. First, the formulation of a state variable model requires explicit recognition of one's assumptions concerning behavioral choices, constraints, dynamics, and so on. Second, the effect of altering such assumptions can be readily assessed by appropriate modification of the model. We have discussed some such modifications above, but many others might also be of interest. Additional field observations may often be required in order to refine the hypotheses underlying a particular model.

For example, our models of lion behavior have assumed that only one type of prey is hunted in any given 30-day period. A more realistic model would allow for several prey types to be available simultaneously, depending on the season. Also, by considering only the objective of survival, our model completely ignores breeding potential. Given data relating the condition of female lions at the end of the nonbreeding season to subsequent breeding success, one could introduce a more realistic terminal fitness function $\phi(X(T))$. This could affect predictions of group size for $t$ near $T$, but would probably not affect predictions for $t \ll T$ (see

Section 2.3). Models involving the care and feeding of young lions could also be developed (see Chapter 6).

Other advantages of living in groups, such as improved defense of kills against scavengers, could also be included. The main purpose of the models developed in this chapter, however, has been to provide quantitative estimates of the contribution to individual fitness that could result from group hunting and communal sharing, for the types of prey and ecological conditions in the Serengeti. In performing this assessment, we have deliberately not "loaded" the models in favor of group behavior.

To summarize, the models developed in this chapter have provided support for Packer's (1986) hypothesis that, under the ecological conditions characteristic of the African savannah, conspecific scavenging within lion prides is highly advantageous to individual lions, and may have underlain the evolution of lion sociality. Certain anthropologists (e.g. Kurland and Beckerman 1985) have suggested independently that the evolution of human social behavior may have been triggered by the same mechanisms.

# 4 Reproduction in Insects

In this chapter we consider the reproductive behavior of insects, such as Hymenoptera that are parasites of other insects, or Tephritid flies that parasitize fruit. We begin in Section 4.1 with a discussion of fitness through egg production. In the rest of the chapter we develop a sequence of state variable models of increasing complexity to describe a variety of host–insect situations.

## 4.1 Fitness from Egg Production and Experimental Background

Most insects encounter hosts, lay eggs, and then continue to look for other hosts. We assume that once eggs are laid, the insect does nothing to add to the fitness that it accrues from the egg, so that the fitness that an insect accrues from oviposition in a host of a certain type depends only upon the kind of host, number of eggs laid, and environmental conditions that are essentially out of the insect's control.

To begin, we consider how egg production affects fitness of the adult female. (We disregard male behavior entirely.) Fitness could be measured by (i) the number of viable offspring produced, (ii) the number of offspring that reach adulthood (a special definition of "viable"), or (iii) the fecundity of offspring (essentially the number of third-generation offspring produced). All three definitions have been used in the literature.

Use of either viable offspring or fecundity of offspring as a measure of fitness also implicitly assumes that fitness is measured one or two generations ahead, rather than many generations ahead. There are a number of reasons for doing this. First, the available data on first-generation offspring are very good, whereas data for succeeding generations are more difficult to obtain and often of lower quality.

Second, by considering the number of grandchildren (through fecundity of offspring), we include everything that a female insect can do to affect her offspring through all generations. (The number of great-grandchildren is determined by the behavior of the daughters of the female we are modeling, etc.)

If we adopt the definition of fitness in terms of the number of viable offspring, then the fitness that a female insect accrues by laying a clutch of size $c$ in a host of type $i$ can be expressed as

$$\delta F = c\rho_i(c) \tag{4.1}$$

where $\delta F$ is the increment in fitness and $\rho_i(c)$ is the probability that an individual egg of the clutch reaches maturity when $c$ eggs are laid on a host of type $i$. On the other hand, if fitness is defined in terms of fecundity of viable offspring and $f_i(c)$ is the expected fecundity of an offspring that reaches maturity when it is one of a clutch of $c$ eggs on a host of type $i$, then Eq. (4.1) is replaced by

$$\delta F = c\rho_i(c)f_i(c). \tag{4.2}$$

It is usually reasonable to assume that the survivorship $\rho_i(c)$ is a decreasing function of $c$. One commonly used model is based on a linear decrease:

$$\rho_i(c) = 1 - \gamma_i c \tag{4.3}$$

where $\gamma_i$ is a measure of the sensitivity of larval survival to crowding. The fitness increment in Eq. (4.1) is then a quadratic function with a maximum at $c_i^* = \gamma_i/2$. One can then think of the clutch $c_i^*$ as the *single host maximum* (SHM) clutch size for hosts of type $i$.

Can insects recognize hosts of various types? The answer varies with the insect, host, and other environmental parameters, but there is clear evidence that host recognition is a common phenomenon (van Alphen 1980, van Alphen and Nell 1982, Courtney 1986, Dethier 1982, Gossard and Jones 1977, Grossmueller and Lederhouse 1985, Hayes 1985, van Lenteren et al. 1978, Rausher 1979, Root and Kareiva 1984).

Considerable variation is observed in the experimental study of clutch size (e.g. Carey 1984, Fitt 1984, and McDonald and McInnis 1985). We want to develop methods to determine if this variation represents an adaptive adjustment of clutch size to host types and other variables.

The models developed in this chapter are motivated by the following experiments:

Table 4.1
Coefficients used in the fitness function $\delta F_i = A_0 + A_1 c + A_2 c^2 + A_3 c^3$ (valid for $c \leq 30$), for *N. vitripennis* (Figure 4.1b).

| host type | approx. average host volume* | $A_0$ | $A_1$ | $A_2$ | $A_3$ | SHM clutch |
|---|---|---|---|---|---|---|
| 1 | 17 | -.2302 | 2.7021 | -.2044 | .0039 | 9 |
| 2 | 22 | -.1444 | 2.2997 | -.1170 | .0013 | 12 |
| 3 | 32 | -.1048 | 2.2097 | -.0878 | .0004222 | 14 |
| 4 | 37 | -.0524 | 2.0394 | -.0339 | -.0003111 | 23 |

*From Charnov and Skinner (1984).

*Variation in clutch size.* Charnov and Skinner (1984, 1985) estimated fitness increments from oviposition of certain wasps.

Figure 4.1 shows the results of their computations of the fitness increment to a mother from oviposition on different hosts for *Trichogramma* (upper panel) and *Nasonia vitripennis* (lower panel). In each case the fitness increment is a concave function of clutch size, depending upon host type. If the curves in Figure 4.1b are extended symmetrically past the maximum, the curves can be fit very well using least squares to a cubic equation of the form

$$\delta F_i(c) = A_0(i) + A_1(i)c + A_2(i)c^2 + A_3(i)c^3 \qquad (4.4)$$

where the parameters $\{A_k(i)\}$ are functions of host type $i$. In using Eq. (4.4), we treat clutch size $c$ as a continuous variable, but in the calculations which follow only integer values of clutch size are allowed. Table 4.1 shows values of the parameters for the data in the lower panel of Figure 4.1 (Derr et al. 1981, Klomp and Teernik 1967, and Mackauer 1982 also provide data that can be used for computing increments in fitness for oviposition on a single host.)

For the *Nasonia* data, the host type is indexed by volume. Thus, the SHM clutch is a function of host volume $V$; it can be denoted by $c^*(V)$. The data in Figure 4.1 and Table 4.1 lead to the approximate relationship $c^*(V) = kV$, where $k$ is a proportionality constant. Charnov and Skinner (1984) measured host volumes and clutch sizes; see Figure 4.2. They found that the the wasps laid clutches of virtually any size less than or equal to the SHM

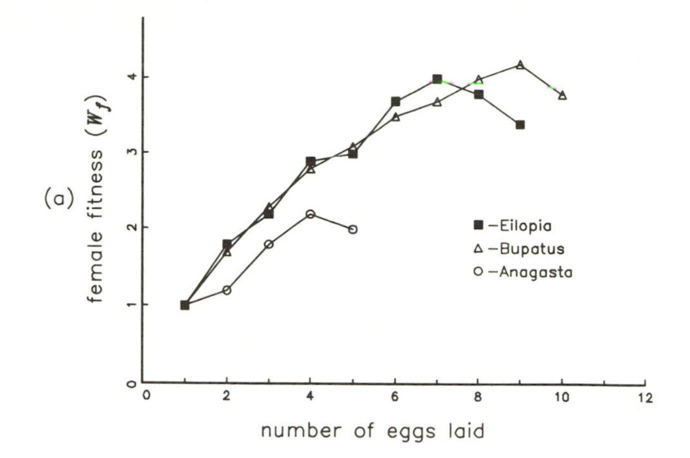


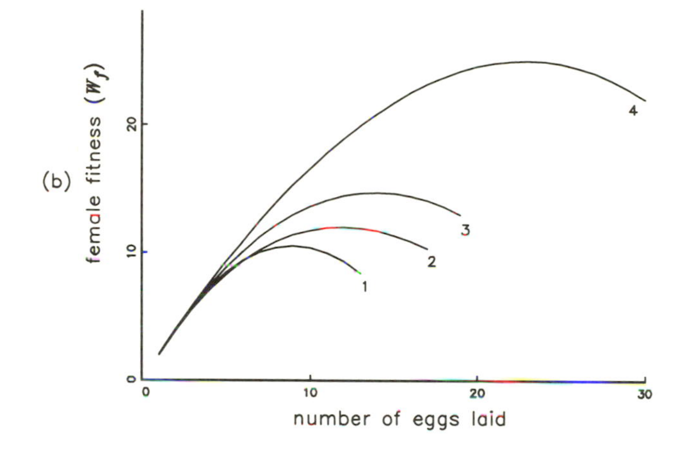


Figure 4.1 Single host fitness increments computed by Charnov and Skinner (1984). (a) Fitness increments for the wasp *Trichogramma* on three different host types. (b) Fitness increments for the wasp *Nasonia* for four hosts of different volumes. (Redrawn from *Florida Entomologist* Vol. 67, 1984.)

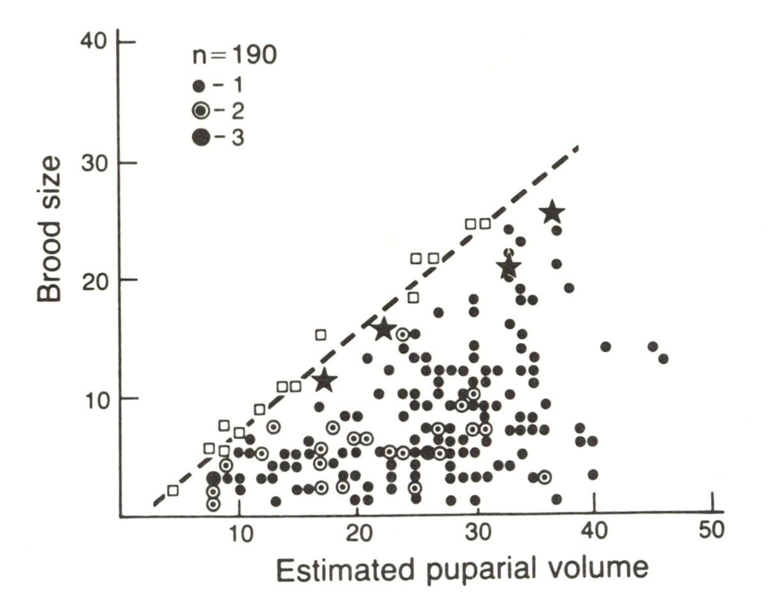


Figure 4.2 Field clutch sizes observed by Charnov and Skinner (1984) for *Nasonia* in hosts of different volumes. The stars denote the SHM clutch. Note that clutches of virtually any size less than the SHM were laid and that small clutches seem to predominate. (Reprinted from *Florida Entomologist* Vol. 67, 1984.)

clutch, but essentially no clutches larger than the SHM clutch. In addition, small clutches predominated. These results are not consistent with the hypothesis that the wasps lay SHM clutches of size $c^*(V)$. Charnov and Skinner (1984, 1985) and Skinner (1985) considered an alternative hypothesis, that clutches smaller than the SHM are observed because the insects are maximizing the rate of total fitness through oviposition per host (determined by both search time for hosts and handling time on hosts), and suggested that this measure of fitness is superior to the SHM measure of fitness. They propose a marginal value type argument (Charnov 1976) which does in fact show that under appropriate conditions the optimal oviposition behavior for an insect is to lay clutches smaller than the SHM. Marginal value arguments do not, however, provide any explanation of the variation in clutch size. Weis et al. (1983) apply demographic ideas involving survivorship and fecundity to their data on the gall maker. They develop a model in which the optimum clutch shifts from one egg per host to three eggs per host as the probability of surviving between host encounters decreases from 1.0 to 0.9. Iwasa et al. (1984) develop a series of models for oviposition behavior using stochastic dynamic programming. Their work is the closest to the models developed in this chapter, but the dynamic programming equations are independent of time.

*Response of apple maggots to host deprivation.* Roitberg and Prokopy (1983) studied the response of the apple maggot *Rhagoletis pomonella* to host deprivation. This fly lays one egg per oviposition and marks the fruit afterwards with an oviposition marking pheromone (OMP). The flies normally avoid ovipositing in pheromone-marked fruit, presumably because the fruit is typically unable to support two larvae and the first larva has a greater chance of survival than the second larva. Roitberg and Prokopy found that after relatively short periods of host deprivation, however, the flies had a considerably increased proclivity to oviposit in marked fruit. For example, about 10% of the flies studied would oviposit in marked fruit when the time elapsed since the last oviposition was 5 minutes. The percentage climbed to about 65% for a 20 minute host deprivation period and reached about 90% after 80 minutes of host deprivation. We will develop dynamic models that show how clutch size and host choice change with host deprivation (see also Mangel 1987a,b).

*Inverse density dependence.* The *density dependence* of parasitism is called *direct* if the fraction of hosts parasitized increases with host density, and is called *inverse* if the fraction of hosts parasitized decreases with host density. A good recent review of density dependent parasitism is found in Hassell (1986). Here we concentrate on a particular set of experiments (Borowicz and Juliano 1986). Our model of these experiments will show two important features. First, we consider the population consequences of individual behavior. Second, we show how the process of dynamic modeling and resulting analysis identifies experimental measurements crucial for understanding a natural system.

## 4.2 A Model with Mature Eggs Only

We begin with a simple oviposition model, in which the state variable $X(t)$ denotes the number of mature eggs remaining in an insect's body at the start of period $t$. We assume that the insect begins its adult life with a full complement of mature eggs and that no eggs are produced during its life. This assumption applies to Lepidoptera (moths and butterflies) in general, some Diptera (flies) and some Coleoptera (beetles). In the next section, more complicated state variable models are described.

The initial egg reserve will be denoted by $X(0) = R$. In light of the assumptions, it is always true that $X(t) \geq X(t+1)$, with equality holding only if no eggs are laid during period $t$. The state variable dynamics are thus:

$$X(t+1) = X(t) - C(t) \tag{4.5}$$

where $C(t) \geq 0$ is the clutch laid during period $t$.

We assume that the insect may encounter one of $H$ different kinds of hosts, and we define

$$\begin{aligned} \lambda_i(t) = \Pr\{&\text{encountering a host of type } i \text{ during period } t\} \\ &i = 1, 2, \ldots, H. \end{aligned} \tag{4.6}$$

In practice, the encounter probability may also depend upon the value of the state variable at the start of period $t$. For example, insects with a small number of eggs remaining may not look as avidly for host sites as would insects of the same age but with a larger complement of eggs. However, we will not consider such situations in this section. The probability of not encountering any host during period $t$ is denoted by

$$\lambda_0(t) = 1 - \sum_{i=1}^{H} \lambda_i(t). \tag{4.7}$$

Finally we assume that the length of periods is sufficiently small that insects encounter at most one host in each period.

The increment in fitness that the insect accrues if it lays a clutch of size $c$ in a host of type $i$ during period $t$ when $X(t) = x$ is denoted by $W_i(c, x, t)$. Survival of the insect between one period and the next is modeled by

$$\begin{aligned} \rho_i(c, x, t) = \Pr\{&\text{insect is alive at the start of period } t+1 \\ &\text{given that it is alive at the start of} \\ &\text{period } t,\ X(t) = x, \text{ and a clutch of} \\ &\text{size } c \text{ is laid in host } i \\ &\text{during period } t\}. \end{aligned} \tag{4.8}$$

Note that the survival function may depend upon the type of host encountered as well as the oviposition decision, e.g. some hosts are more dangerous than others. The probability of surviving during period $t$ if no host is encountered will be denoted by $\rho_0(t)$. The survival functions can also be determined by direct experimental measurement.

The time horizon $T$ may have different interpretations: (i) it may be the time at which the insect is dead with probability 1; (ii) it may be the time at which hatch probability drops to 0 (this usually occurs substantially before death—see Carey 1984 for an example); (iii) it may be the end of a foraging day, for an insect whose unused eggs are resorbed overnight.

We now define the expected lifetime fitness function $F(x,t,T)$ as

$$F(x,t,T) = \text{maximum expected lifetime fitness obtained through egg production, between } t \text{ and } T, \text{ given that } X(t) = x. \tag{4.9}$$

The dynamic programming equation for $F(x,t,T)$ is derived in a manner analogous to the derivation of Eq. (2.10), except that the decision is no longer about patch choice. Instead, if the insect encounters a host of type $i$ in a given period, then the decision is how large the clutch should be. The clutch laid if a host is encountered during period $t$ is bounded by 0 as the minimum and $X(t)$ as the maximum. The dynamic programming equation for $F(x,t,T)$ thus becomes:

$$\begin{aligned} F(x,t,T) &= \lambda_0(t)\rho_0(t)F(x,t+1,T) \\ &+ \sum_{i=1}^{H} \lambda_i(t) \max_{0 \le c \le x} \{W_i(c,x,t) + \rho(c,x,t)F(x-c,t+1,T)\} \\ &\text{for } x > 0 \end{aligned} \tag{4.10}$$

and $F(x,t,T) = 0$ for $x = 0$. The terms on the right-hand side of Eq. (4.10) are interpreted as follows. The first term corresponds to the situation in which no hosts are encountered in period $t$, so that the state variable does not change. The summation is taken over all host types; $\lambda_i$ is the probability that a host of type $i$ is encountered during period $t$. When a host is encountered, the decision is the size of the clutch $c$, constrained by the number of eggs remaining. The first term in brackets is the immediate increment in fitness and the second term is the future expected lifetime fitness associated with the reduced number of remaining eggs. The end condition is $F(x,T,T) = 0$, so that there is no fitness associated with having any eggs remaining at the beginning of period $T$.

The algorithm for solving Eq. (4.10) is a modest modification of the patch selection algorithm given in Chapter 2. Before discussing results obtained by solving the dynamic programming equation, however, it is worthwhile to consider the intuition that can be obtained by studying the equation itself.

For example, consider the case in which $t = T - 1$. Then Eq. (4.10) becomes

$$F(x,T-1,T) = \sum_{i=0}^{H} \lambda_i(T-1) \max_{0 \le c \le x} \{W_i(c,x,T-1)\}. \quad (4.11)$$

When $t = T - 1$, there is no fitness associated with future oviposition. If a host of type $i$ is encountered, the number of eggs laid in this host should be the smaller of the SHM clutch or the number of eggs remaining. (There is no reason ever to lay more eggs than the SHM because fitness from this clutch would be reduced without any future benefit.) When $t$ is far from $T$, on the other hand, Eq. (4.10) shows that if a host of type $i$ is encountered, the immediate fitness increment obtained from oviposition in this host should be balanced against the future expected fitness from all hosts. In such a case, it is likely that a host for which the fitness increment is small will receive no eggs at all; that is, the optimal oviposition decision will be $C(t) = 0$. This reasoning leads to two immediate predictions that can be tested in either field or laboratory studies:

*Prediction 1:* For a fixed number of remaining eggs, older insects should lay larger clutches than younger insects. In addition hosts that are inferior and thus not chosen by young insects will be acceptable to older insects (depending on their egg complements).

*Prediction 2:* Imagine a cohort of insects that emerge at the same time with exactly the same initial complement of eggs. At some later time, as a result of differing host encounters, there will be a distribution in the number of eggs remaining per insect. This will lead to a distribution in the clutch sizes laid by the cohort. Conversely, identical insects kept in tightly controlled laboratory situations, so that the distribution in egg complement is relatively controlled, should lay clutches of essentially the same size at any given time.

Notice that Prediction 2 implies that the data obtained from an oviposition experiment in which egg complement is not controlled will consist of a scatter of clutch sizes, all less than or equal to the SHM; cf. Figure 4.2.

An additional "analytic" observation is the following: The survival function $\rho_i(c, x, t)$ acts to "discount" the value of future ovipositions. That is, as the survival function decreases, the relative value of future expected fitness through oviposition decreases when compared to the fitness increment from oviposition in the current host. This observation leads to another testable prediction:

*Prediction 3:* As the per period survival probability decreases, larger clutches will be observed.

In order to test this prediction, one needs to learn the cues that the insects have evolved to indicate when shifts in mortality are likely. An analogous idea would be experiments in which fish are presented with model predators (e.g. Milinski 1986, Pitcher 1986). There is anecdotal evidence (B. Roitberg, personal communication) that some *Rhagoletis* flies alter their oviposition behavior in response to pressure changes which indicate an impending rain storm, and thus a higher future mortality. The development of good experiments to test predictions analogous to 3 is a promising area of future research.

In order to solve the dynamic programming equation, the fitness increment and survival functions must be given explicitly. In the results reported below, the following two survival functions are used:

$$\text{(a)} \qquad \rho(c, x, t) = \begin{cases} 1 & \text{for } t < T - 1 \\ 0 & \text{for } t = T - 1 \end{cases} \tag{4.12}$$

$$\text{(b)} \qquad \rho(c, x, t) = l_{t+1}/l_t$$

where

$$l_t = 1 - (t/T)^\gamma \tag{4.13}$$

where $\gamma$ is a parameter. The first choice of survival function is the simplest that one could imagine: 100% survivorship through every period with certain mortality at the end of the interval $[0, T]$. The second choice corresponds to survival depending only upon time and independent of egg complement or decision. The function $l_t$ can be interpreted as the probability of surviving to period $t$ from period 1. The choice of the parameter $\gamma$ determines the shape of the survival curve $l_t$. For $\gamma = 1$ we obtain a line. As the value of $\gamma$ increases above 1, $l_t$ assumes the following shape: for small values of $t$, $l_t$ is approximately 1, with a sudden decrease as $t$ increases and approaches $T$.

When the dynamic programming equation, Eq. (4.10), is solved, we obtain $F(x, t, T)$, i.e. the maximum expected fitness through egg production between $t$ and $T$ and, more importantly, we also obtain the optimal oviposition decisions as a function of time to go and current egg complement, for each type of host that may be encountered. The optimal decision is, in many ways, more interesting than the fitness function itself because clutch sizes are what will be measured in field or laboratory experiments. For example, Table 4.2 shows the optimal oviposition decisions for the fitness increments given in Table 4.1 when a host of either type 2 (SHM clutch = 12) or type 4 (SHM clutch = 23) is encountered, for $X(t) = 40$ or 80 eggs, as a function of time to go $T - t$. (Other parameter values are listed at the bottom of Table 4.2.) The results presented in this table show that when the time that remains for foraging for oviposition sites is small, the optimal decision is to lay the SHM clutch, but that as the time remaining increases, smaller clutches become optimal. The source of this behavior is the concavity of the fitness increment function $W_i(c, x, t)$. That is, if survivorship were guaranteed and the time to go were sufficiently long, then a concave fitness function would lead to the insect placing as few eggs as possible in each host. (The conceivable lower limit here is 1 egg per host.) The optimality of clutches with a larger number of eggs per host is a result of the interplay of the time horizon, mortality risk, and egg load.

The results presented in Table 4.2 do not completely explain the variability in clutch sizes observed in the experiments. However, this variability can be explained by following the probability distribution of the state variable $X(t)$ over time. Given the probability distribution of the state variable, we can predict the distribution of clutch sizes in each period. One way of comput-

Table 4.2
Optimal oviposition decisions*

| Time to go | Optimal Oviposition Decisions | | | |
|---|---|---|---|---|
| | Host Type 2 | | Host Type 4 | |
| $(T-t)$ | $x=40$ | $x=80$ | $x=40$ | $x=80$ |
| 1 | 12 | 12 | 23 | 23 |
| 2 | 12 | 12 | 20 | 23 |
| 3 | 9 | 12 | 14 | 23 |
| 4 | 6 | 12 | 11 | 21 |
| 5 | 5 | 10 | 9 | 17 |
| 6 | 4 | 8 | 8 | 15 |
| 7 | 4 | 7 | 6 | 13 |
| 8 | 3 | 6 | 6 | 11 |
| 9 | 3 | 5 | 5 | 10 |
| 10 | 3 | 5 | 5 | 9 |
| 11 | 2 | 4 | 4 | 8 |
| 12 | 2 | 4 | 4 | 8 |
| 13 | 2 | 4 | 4 | 7 |
| 14 | 2 | 3 | 3 | 6 |
| 15 | 2 | 3 | 3 | 6 |
| 16 | 2 | 3 | 3 | 6 |
| 17 | 2 | 3 | 3 | 5 |
| 18 | 2 | 3 | 3 | 5 |
| 19 | 2 | 3 | 3 | 5 |
| 20 | 2 | 3 | 2 | 5 |

* Survival function given by Eq. (4.12a) and $\lambda_1 = .05$, $\lambda_2 = .05$, $\lambda_3 = .1$, $\lambda_4 = .8$.

ing the probability distribution of $X(t)$ is by the use of forward iteration as described in Appendix 2.3. However, we will here use the method of *Monte Carlo simulation.* The simulation approach is particularly useful for problems in which the number of state variables is large. For example, if host densities change over time then one needs many state variables to describe the system: egg complement $X(t)$ and one state variable for each host density or encounter probability. In such a case, the analytical approach described in Appendix 2.3 is unwieldy and the Monte Carlo approach described below is much easier to use.

In order to use the Monte Carlo approach, a random number generator is required. Most microcomputers have internal random number generators which produce uniformly distributed random numbers. These are often sufficient.* To develop the algorithm, we introduce a "cohort" of model insects. To do this, let $X_j(t)$ denote the egg complement of the $j$th insect in the cohort at the start of period $t$ for $j = 1, 2, \ldots J$, where $J$ is the total number of model insects followed in the simulation. The simulation procedure keeps track of the host encounters, oviposition decisions, and the egg complement of each individual insect. Each insect corresponds to one independent trial in the Monte Carlo simulation, so that in order to obtain a representative sample we want to choose the value of $J$ sufficiently large. Values in the range of $10 \le J \le 100$ typically are sufficient. The algorithm for the problem corresponding to Eq. (4.10) proceeds as follows:

1. *Initialization step.* The first step is to store in the computer the parameters $\lambda_i$, for $i = 1$ to 4, the optimal decision matrix $c^*(x, t, i)$ (which is the optimal clutch for a host of type $i$ encountered in period $t$ when $X(t) = x$) and the initial egg complement of each insect. For example, if we assume that each of $J$ insects started with the same egg complement, then $X_j(1) = R$ for $j = 1, 2, \ldots, J$. Time is initialized by setting $t = 1$.

2. *Encounter step.* Increment $t$ by 1 unit, so that $t$ is replaced by $t + 1$. For each insect, use the random number generator to determine the kind of host encountered. To do this, cycle through $j$ and implement the following procedure.

   (a) Pick a random number $Z$ distributed uniformly on $[0, 1]$, using the random number generator.

   (b) Determine which type of host is encountered: for the problem described in this section, with 4 host types, proceed as follows. If $0 \le Z < \lambda_1$ then a host of type 1 is encountered

* We should warn the reader that some random number generators built in to microcomputers are not very reliable: see *Byte*, Vol. 12, No. 1 (Jan. 1987), p. 175.

in the current period $t$. If $\lambda_1 \leq Z < \lambda_1 + \lambda_2$ then a host of type 2 is encountered, if $\lambda_1 + \lambda_2 \leq Z < \lambda_1 + \lambda_2 + \lambda_3$ then a host of type 3 is encountered, and so on. Finally, if $Z \geq \sum \lambda_i$ then no host is encountered in the current period $t$.

(c) Determine the clutch size: If a host of type $i$ is encountered by the $j$th insect during period $t$, then the clutch laid by that insect is $c^*(X_j(t), t, i)$.

3. *Output step.* For each insect, print the clutch laid during period $t$. (Or you may wish to store the data in some other appropriate way, outputting only summary statistics.) If $t < T$, update the insect's egg complement, and return to Step 2.

The simulation algorithm is easy to implement and can be used to perform "experiments" on "computer insects" by varying particular parameters such as encounter rate probabilities or time horizon.

We now describe two sets of computer "experiments" using the simulation approach. The first experiment studies the effect of host distribution, determined by the values of the encounter probabilities, on the distribution of clutch sizes. This is meant to mimic the observations of Charnov and Skinner (1984). The second experiment studies the effect of time horizon on the clutch size distribution. This is motivated by the experiments of Roitberg and Prokopy (1983).

Table 4.3 shows the various combinations of time horizon, encounter probabilities, and survival function used in the experiments. The results of the host distribution experiment are shown in Table 4.4. These experiments show the same kind of variability in clutch size distribution that Charnov and Skinner (1984) observed. In particular, small clutches predominate and the SHM clutch occurs infrequently. The computer generated data also confirm the qualitative predictions made by study of the dynamic programming equation.

Table 4.5 shows the effect of time horizon on clutch size. There is a clearly pronounced shift towards larger clutches as the time available for foraging for oviposition sites decreases. This result is analogous to the experiments of Roitberg and Prokopy (1983) in which *Rhagoletis pomonella* flies oviposited more readily in pheromone-marked fruit after periods of host deprivation. A more detailed model of these experiments is given in Mangel (1987b).

Table 4.3
Parameter values used in different cases for the computer experiments

| Case | $T$ | $\lambda_1$ | $\lambda_2$ | $\lambda_3$ | $\lambda_4$ | Survival Function |
|---|---|---|---|---|---|---|
| 1 | 20 | .1 | .1 | .1 | .1 | 4.12(a) |
| 2 | 20 | .05 | .05 | .1 | .8 | 4.12(a) |
| 3 | 15 | .1 | .1 | .1 | .1 | 4.12(a) |
| 4 | 15 | .05 | .05 | .1 | .8 | 4.12(a) |
| 5 | 10 | .1 | .1 | .1 | .1 | 4.12(a) |
| 6 | 10 | .05 | .05 | .1 | .8 | 4.12(a) |
| 7 | 20 | .1 | .1 | .1 | .1 | 4.12(b), $\gamma = 1$ |
| 8 | 20 | .05 | .05 | .1 | .8 | 4.12(b), $\gamma = 1$ |
| 9 | 15 | .1 | .1 | .1 | .8 | 4.12(b), $\gamma = 1$ |
| 10 | 15 | .05 | .05 | .1 | .8 | 4.12(b), $\gamma = 1$ |
| 11 | 10 | .1 | .1 | .1 | .1 | 4.12(b), $\gamma = 1$ |
| 12 | 10 | .05 | .05 | .1 | .8 | 4.12(b), $\gamma = 1$ |
| 13 | 20 | .1 | .1 | .1 | .1 | 4.12(b), $\gamma = 2$ |
| 14 | 20 | .05 | .05 | .1 | .8 | 4.12(b), $\gamma = 2$ |
| 15 | 15 | .1 | .1 | .1 | .1 | 4.12(b), $\gamma = 2$ |
| 16 | 15 | .05 | .05 | .1 | .8 | 4.12(b), $\gamma = 2$ |
| 17 | 10 | .1 | .1 | .1 | .1 | 4.12(b), $\gamma = 2$ |
| 18 | 10 | .05 | .05 | .1 | .8 | 4.12(b), $\gamma = 2$ |

Table 4.4
Results of the host distribution experiment. (Entries are frequency of observed clutches.*)

| | Clutch Size | | | | |
|---|---|---|---|---|---|
| Case | 1–5 | 6–10 | 11–15 | 16–20 | 21–25 |
| 1 | .02 | .35 | .41 | .15 | .07 |
| 2 | .24 | .76 | 0 | 0 | 0 |
| 7 | .01 | .26 | .52 | .04 | .17 |
| 8 | .14 | .29 | .57 | 0 | 0 |
| 13 | .01 | .27 | .51 | .05 | .15 |
| 14 | .16 | .36 | .48 | 0 | 0 |

*Cases 1,7, and 13 correspond to all $\lambda_i = .1$. Cases 2, 8, and 14 correspond to $\lambda_1 = \lambda_2 = 0.05$, $\lambda_3 = 0.1$, $\lambda_4 = 0.8$. In all cases, $T = 20$. The survival function is Eq. (4.11a) for cases 1, 2; Eq. (4.11b) with $\gamma = 1$ for cases 7 and 8, and Eq. (4.11b) with $\gamma = 2$ for cases 13, 14.

Table 4.5

Results of the host distribution experiment. (Entries are the frequencies of observed clutches.*)

| Case | $T$ | Clutch Size 1–5 | 6–10 | 11–15 | 16–20 | 21–25 |
|---|---|---|---|---|---|---|
| 1 | 20 | .02 | .35 | .41 | .15 | .07 |
| 3 | 15 | .01 | .30 | .46 | .05 | .19 |
| 5 | 10 | 0 | .26 | .48 | .003 | .26 |
| 2 | 20 | .24 | .76 | 0 | 0 | 0 |
| 4 | 15 | .16 | .82 | .02 | 0 | 0 |
| 6 | 10 | .02 | .21 | .74 | .02 | .002 |
| 7 | 20 | .01 | .26 | .52 | .04 | .17 |
| 9 | 15 | .004 | .25 | .50 | .01 | .25 |
| 11 | 10 | 0 | .23 | .51 | 0 | .26 |
| 8 | 20 | .14 | .29 | .57 | 0 | 0 |
| 10 | 15 | .08 | .28 | .33 | .32 | 0 |
| 12 | 10 | .03 | .18 | .14 | .66 | 0 |
| 13 | 20 | .013 | .27 | .51 | .05 | .15 |
| 15 | 15 | .003 | .24 | .49 | .02 | .25 |
| 17 | 10 | 0 | .24 | .51 | 0 | .25 |
| 14 | 20 | .16 | .36 | .48 | 0 | 0 |
| 16 | 15 | .11 | .27 | .62 | 0 | 0 |
| 18 | 10 | .04 | .19 | .18 | .59 | 0 |

*Cases 1–6 have survival function Eq. (4.12a). Cases 7–12 have survival function Eq. (4.12b), with $\gamma = 1$. Cases 13–18 have survival function Eq. (4.12b) with $\gamma = 2$.

For odd numbered cases, all $\lambda_i = 0.1$; for even numbered cases $\lambda_1 = \lambda_2 = 0.05$, $\lambda_3 = 0.1$ and $\lambda_4 = 0.8$

## 4.3 A Model with Mature Eggs and Oocytes

We next briefly describe a model in which the insect matures eggs from oocytes. There are now two state variables: $X(t)$ denotes the number of mature eggs at the start of period $t$ and $Y(t)$ denotes the number of oocytes remaining at the start of period $t$. In addition to the depletion of mature eggs through oviposition, there is a transfer of oocytes to mature eggs and a corresponding depletion of oocytes. (A more complex model is one in which oocytes can be replenished by a search for food or other nutrient resources. Such a model is an easy extension of the ones described in this and the previous section.)

To model the state variable dynamics, one needs to introduce an "egg production function" denoted by $r(x, y, t)$ and defined as follows:

$$r(x, y, t) = \text{number of oocytes matured into eggs during period } t, \text{ given that } X(t) = x,\ Y(t) = y. \tag{4.14}$$

In principle, this production function can be determined by dissection of insects in the laboratory. Some of its properties are clear. For example, $r(x, 0, t) = 0$ regardless of the value of $x$ or $t$. If the insect can hold no more than $X_m$ mature eggs, then $r(x, y, t) = 0$ if $x > X_m$.

Assume that the temporal sequence is first oviposit, then mature eggs. Then the dynamics for the state variables are

$$\begin{aligned} X(t+1) &= X(t) + r(X(t) - C(t), Y(t), t) - C(t) \\ Y(t+1) &= Y(t) - r(X(t) - C(t), Y(t), t). \end{aligned} \tag{4.15}$$

Here $C(t)$, the clutch laid during period $t$, is a decision variable which depends upon which host the insect encounters. The lifetime fitness function is denoted by $F(x, y, t, T)$:

$$F(x, y, t, T) = \text{maximum expected lifetime fitness obtained through egg production between } t \text{ and } T, \text{ given that } X(t) = x \text{ and } Y(t) = y. \tag{4.16}$$

The dynamic programming equation for $F(x, y, t, T)$ is easy to derive; we encourage the reader to do it as an exercise. There are essentially no conceptual difficulties with using this more complicated model, and the computational difficulties are relatively minor as well. It is likely that the biggest problem with a small computer will be a shortage of memory space, limiting the size of the fitness matrix that can be stored in the computer.

## 4.4 Parasitism and Density Dependence

We now develop a model, motivated by the work of Borowicz and Juliano (1986) on a fruit fly, in which the density dependence of host parasitism can be studied. One of our aims is to show how the individual behaviors predicted with the dynamic modeling approach can be connected to population level phenomena. A second aim is to show how dynamic modeling leads to insights about experimental variables that must be measured.

In the experiments, host densities change over time, as a result of parasitism by the flies. Thus another state variable must be considered, namely the distribution of host types over time. In the models that follow, we will consider four host types. Two of these are unparasitized "clean" hosts and two are hosts that have been previously parasitized ("marked" with a pheromone), each providing a different level of fitness. In the experiments of Borowicz and Juliano, the clean hosts were termed blue (mature) and brown (immature). We will simply call them type 1 and type 2 and assume that fitness accrued to a fly through oviposition in a host of type 1 has a relative value equal to 1, and fitness through oviposition in a host of type 2 has a relative value of $\gamma < 1$. Host types 3 and 4 will respectively denote marked blue and brown hosts with fitnesses $p$ and $\gamma p$ where $p < 1$. If $N_i(t)$ denotes the number of hosts of type $i$ at the start of period $t$, then $\sum N_i(t)$ always equals $N_1(0) + N_2(0) = N_0$, but the values of each $N_i(t)$ change over time in response to fly oviposition decisions.

One of the major difficulties in relating individual behavior to its population consequences is that the actions of one individual affect the environment experienced by other individuals. The most rigorous way to treat this interaction is by means of game theory; see Chapter 9. Here we finesse the difficulties of dynamic game theory (which are manifold) in the following way. We assume that the flies do not consider the behavior of other individuals when making their own oviposition decisions. Some species of parastic wasps (e.g. *Trichogramma evanescens*) apparently do not change their behavior in response to the presence of other wasps (Waage and Lane 1984). In addition, we assume that the flies assess the distribution of host types at the start of each period and make oviposition decisions solely on the basis of their egg complement and the host distribution at the start of period t. That is, we assume that the flies use past experience and decisions to assess the current state of the environment but do not "forecast" future states.

We model encounters of flies with hosts as a random search process (Mangel 1985a) with parameter $\epsilon$, so that if the distribution of host types at the start of period $t$ is $\{N_i(t)\}$, then the probability of encountering a host of type $i$ during period $t$ is

$$\lambda_i(t) = (1 - \exp(-\epsilon N_0))N_i(t)/N_0. \tag{4.17}$$

The expression $\lambda_i(t)$ in Eq. (4.17) is interpreted as follows. First assume that search for hosts is a Poisson process with encounter rate $\epsilon$ per host (Appendix 1.1). The encounter rate for $N_0$ randomly distributed hosts is then $\epsilon N_0$. The probability of encountering no host in a unit time interval is $\exp(-\epsilon N_0)$, so that the probability of encountering at least one host is $1 - \exp(-\epsilon N_0)$. We assume that at most one host is encountered per period (the fly must also decide whether to oviposit). Given that a host is encountered, the probability that the host is of type $i$ is $N_i(t)/N_0$. We also introduce

$$\lambda_0 = 1 - \sum_i \lambda_i = \exp(-\epsilon N_0) \tag{4.18}$$

as the probability of not encountering any host in period $t$. We assume that the flies forage for oviposition sites for a maximum of $T$ periods and let $\rho(t)$ denote the survivorship between periods. If a host is encountered in a given period, the oviposition decision is either to lay an egg or not to oviposit at all. Let $f_i$ denote the fitness increment from ovipositing in a host of type $i$. If $F(x,t,T)$ denotes maximum expected total lifetime fitness through oviposition between $t$ and $T$ given that $X(t) = x$, then the dynamic programming equation for $F(x,t,T)$ is

$$F(x,t,T) = \lambda_0 \rho(t) F(x,t+1,T) + \sum_{i=1}^{4} \lambda_i(t) \max[f_i + \rho(t)F(x-1,t+1,T); \rho(t)F(x,t+1,T)] \quad (4.19)$$

The first term on the right-hand side of Eq. (4.19) represents the expected fitness from period $t+1$ on if no host is encountered in period $t$, while the terms in the summation represent the maximum expected fitness if a host of type $i$ is encountered. The two expressions following the max correspond to ovipositing or not ovipositing, respectively. The end condition on $F(x,t,T)$ is that no fitness can be accrued after period $T$ so that

$$F(x,T,T) = 0. \quad (4.20)$$

The solution of the dynamic programming equation can be used in Monte Carlo simulations to perform computer experiments analagous to the field experiments of Borowicz and Juliano. The values of the parameters used in the solution and simulation are: $\epsilon = 0.001$, $T = 20$, $p = 0.5$, $\gamma = 0.3$, for all $t$, and an initial egg complement of 10 eggs, and $\rho(t) = 0.99$. Twenty Monte Carlo experiments were performed. The initial fruit density is determined by the values of $N_1(0) = n_1$ and $N_2(0) = n_2$. In all experiments, these were set equal to each other, and denoted by $n_0$. The range of $n_0$ was 10 to 300. Figure 4.3 shows the fraction of fruit parasitized as a function of the value of $n_0$. Two cases are considered in this figure. In Figure 4.3(a) fly density is not adjusted in response to decreasing host density. That is, although $n_0$ varies between 10 and 300, the number of flies is held constant at 10 individuals. One sees a clear "inverse density dependence" in the parasitism. The interpretation is straightforward. When the fruit density is very large, since the flies have a limited egg capacity (a maximum of 10 ovipositions in a given day), the flies can be selective concerning the fruit they choose for oviposition. As the fruit density decreases, the flies become less selective. Since the number of flies is fixed, however, as fruit density decreases the same number of flies are searching for fewer fruits. Fruits which are unacceptable at high densities become acceptable at lower densities and the overall fraction of fruit parasitized increases.

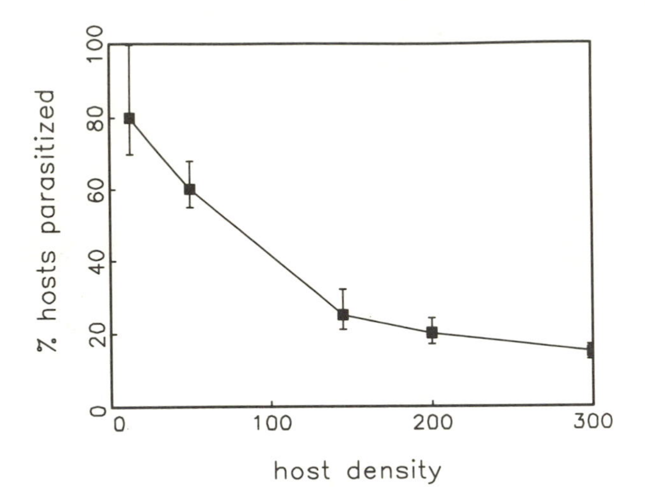


Figure 4.3(a) Fraction of hosts parasitized as a function of host density for the case in which the number of flies is held constant as the host density decreases. In this and the next figure, the bars show the range of fraction of hosts parasitized observed in the Monte Carlo simulations. Note the "inverse density dependence" in the parasitism.

This result is not surprising. As long as the number of eggs is fixed, when the number of fruit increases the relative (percent) parasitism must decrease. All that the state variable approach has done is highlight the need to consider the egg complement of the flies when analyzing parasitism.

Suppose, on the other hand, that hosts are patchily distributed, and that the flies exhibit a numerical response (Holling 1965) to local host density. For example, *Rhagoletis pomonella* appears to respond using visual clues to changes in fruit density (R. Prokopy, personal communication). We thus might assume that the ratio of fruit to flies is constant, rather than the number of flies being constant. Figure 4.3(b) shows the results of Monte Carlo behavioral simulations in which the ratio of fruit to flies is held constant at 10:1. In this case, the fraction of fruit parasitized levels off at high densities, but decreases considerably at low fruit densities. When $n_0 = 10$, there is only one fly searching for the fruit and the chance that it does not find the fruit is considerable (when $n_0 = 10$, the probability of finding a fruit of any kind during one period is only about 0.18). Thus, the limited search abilities of the fly lead to

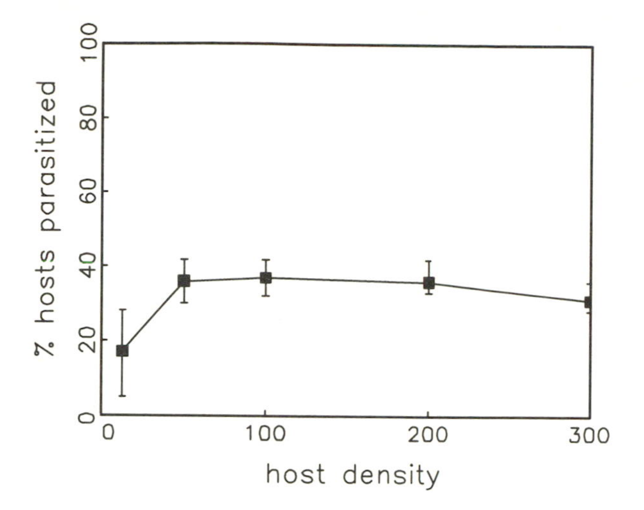


Figure 4.3(b) Fraction of hosts parasitized as a function of host density for the case in which the ratio of hosts to flies is held constant. Note the "direct density dependence" in the parasitism.

a "direct density dependence" of the parasitism at low densities. Borowicz and Juliano (1986) did not report measurements of fly densities or the response of fly density to host density, so that it is not possible to determine which of Figures 4.3(a) or 4.3(b) is appropriate to their experiments. Dynamic behavioral modeling thus indicates key variables that need to be measured in order to understand completely the relationship between host density and fraction of hosts parasitized.

The above model could be extended in various ways. For example, alternating days and nights could be modeled by using sequential coupling (Appendix 2.1.5). Multiple patches of hosts, and patch choice could also be explicitly modeled; such models could also consider informational problems regarding host density in each patch (see Section 9.1). Although such embellishments might be interesting, we feel that the key insight obtained from this model is that in order to understand the density dependence of parasitism, we must have information on both host density and parasite behavior—host data alone are insufficient. The process of formulating and analyzing the dynamic model indicates a key additional experimental measurement.

## 4.5 Discussion

In this chapter, we have shown how the use of dynamic modeling can lead to further understanding of experiments that involve reproduction in insects. Development and analysis of these models of oviposition behavior have indicated the importance of understanding the fitness increment and its effects on lifetime fitness. We have also shown how population consequences of individual behavior can be modeled, and how the formulation of a model can suggest new experiments and key measurements. We believe that the approach of dynamic modeling could lead to new insights when applied to other insect-related problems such as sex ratio allocation (Charnov 1982, Karlin and Lessard 1986), swarming in social insects (Brian 1983, Oster and Wilson 1978), the evolution of pheromone marking systems, or the interaction of flowers and their pollinators (Real 1983).

All of the models developed in this chapter were based on the assumption that the insects could perfectly assess their environment. In particular, we assumed that the insects knew the values of the $\lambda_i$ in the dynamic programming equations. It is most likely, however, that insects must somehow estimate the values of the $\lambda_i$ through host encounter rates. This is a problem in the analysis of information and is deferred until Chapter 9.

# 5 Migrations of Aquatic Organisms

Biologists studying the spatial distribution of fish and their prey have long known that many species of aquatic animals undergo regular daily vertical movements in the water column (Baker 1978, De Courcey 1976, Zaret 1980). These movements are often tied closely to the 24-hour diel cycle, although lunar rhythms may also be involved (Gliwicz 1986b). The typical pattern, called normal vertical migration, is for organisms to rise towards the surface at night and to descend to deeper waters during the day. The opposite pattern, called reverse vertical migration, in which organisms occupy surface waters only during the day, is less common. Finally, some organisms undergo little if any vertical movement. Figure 5.1 shows vertical migration data of seven species of zooplankton, all from a single lake; all three patterns are evident in these data (Narver 1970). The possible evolutionary mechanisms underlying these particular patterns of migration will be discussed in Section 5.3.

It is generally agreed that the proximate cause of normal vertical migration is the diel cycle of changes in light intensity at various depths. Normal vertical migrants tend to follow contours of constant light intensity, at least approximately (Blaxter 1974). Abnormal light conditions, such as those resulting from heavy fog or cloud cover, solar eclipses, bright moonlight, or artificial lights, often alter normal migration patterns.

The ultimate cause of vertical migration is less well understood: "The literature on the vertical migration of zooplankton is enormous, complex, and full of conflicting observations and generalizations" (McLaren 1963, p. 686). Among the hypotheses that have been proposed to explain the adaptive significance of vertical migration are the following:

(1) a method of population self-regulation (Wynne-Edwards 1962);
(2) a way to prevent overexploitation of prey (Hardy 1958);
(3) a mechanism for maximizing the rate of genetic exchange (David 1961);

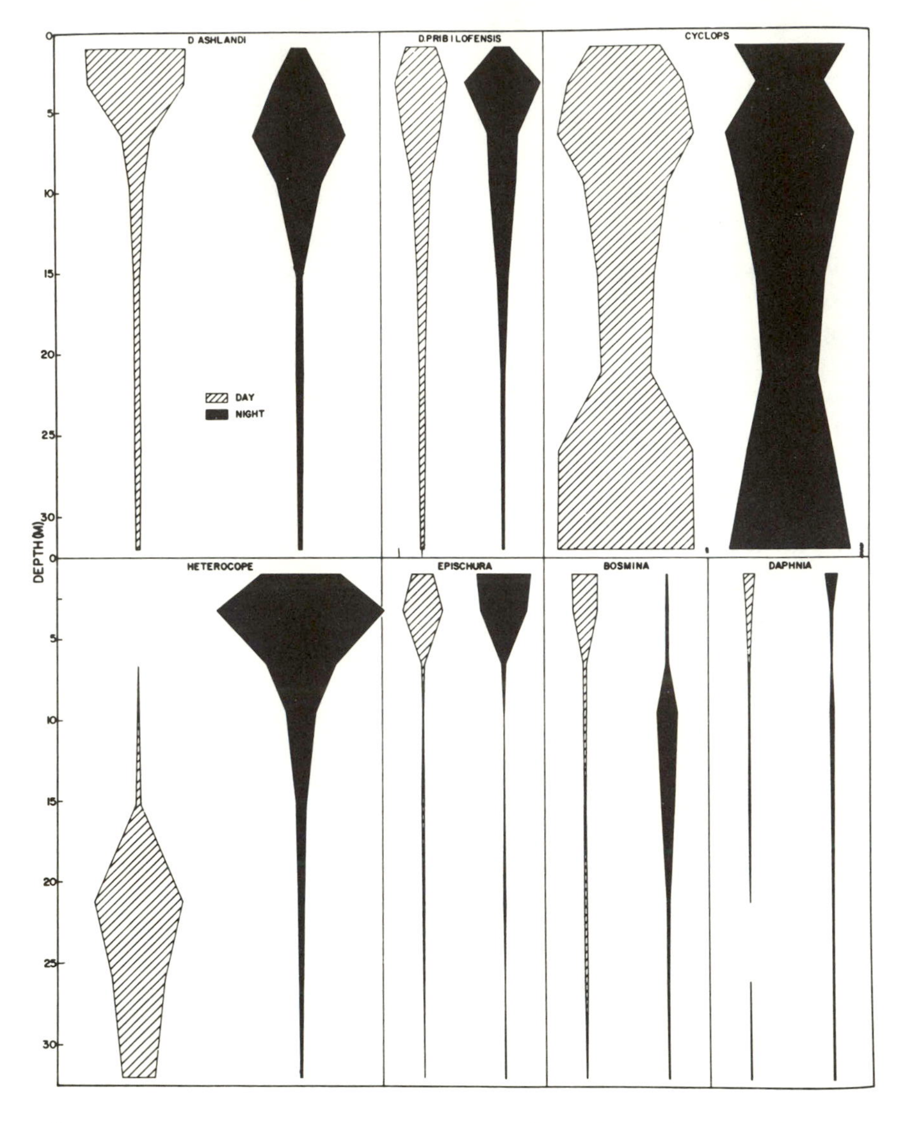


Figure 5.1 Day and night vertical profiles of relative abundance of the major species of zooplankton in Babine Lake, British Columbia, August 5–6, 1967 (from Narver 1970).

(4) a way to avoid harmful solar radiation (Hairston 1976);
(5) a method of maintaining relatively constant food input throughout the year (Kerfoot 1970);
(6) a response to phytoplankton toxicity (Hardy and Gunther 1935);
(7) an effective way of achieving horizontal displacement (Mackintosh 1937);
(8) a response to prey movement or patchiness (Hardy 1958);
(9) a method of minimizing the effects of competition (Dumont 1972, Lane 1975);
(10) a predator avoidance mechanism (Zaret and Suffern 1976);
(11) a response to physical conditions, having no direct biological significance (Hardy 1953);
(12) a method to achieve the optimal alternation of temperature (Moore and Corwin 1956);
(13) an adaptation to bioenergetic requirements (McLaren 1974).

Some of these suggestions are of dubious value, since they rely upon group selection arguments (#1, 2, 3, 5, 9). Several of the suggested hypotheses do relate to individual fitness, but are incapable of predicting the timing of vertical migrations (#6, 7, 12, 13; see Enright 1977). According to these hypotheses, light intensity itself has little or nothing to do with the adaptive significance of vertical migration, but merely acts as an incidental cue.

In many cases several adaptive influences may be involved simultaneously. Current opinion seems to be converging on explanations pertaining to the tradeoff between potential feeding rates and the risk of predation (most aquatic organisms are simultaneously predator and prey), as influenced by light intensity (Hall et al. 1979, Zaret 1980, Iwasa 1982, Cerri 1983, Wurtsbaugh and Li 1985, Gliwicz 1986a,b, Clark and Levy 1988). This is the approach that will be followed in the present chapter.

In Section 5.1 we describe dynamic models of diel vertical migrations of zooplankton. Next, in Section 5.2 we discuss the vertical migrations of planktivores; here a somewhat more subtle treatment of the effects of light intensity is required.*

* In shallow waters horizontal migration between protected littoral or reef zones and open waters (Hall et al. 1979, McFarland et al. 1979, Wurtsbaugh and Li 1985) may have adaptive value similar to that of vertical migration in pelagic areas.

## 5.1 Diel Vertical Migrations of Zooplankton

Babine Lake (British Columbia) contains eight main species of zooplankton, only two of which display pronounced vertical migrations (Figure 5.1). The relatively large predatory calanoid copepod *Heterocope septentrionalis* (approx. 3 mm length) performs normal vertical migrations, rising to the surface at night and descending to deeper water during the day. The cladoceran *Bosmina coregoni*, on the other hand, performs reverse vertical migrations. Narver (1970, p. 313) states that these movements "are rather precisely timed and are probably also related to changes in the intensity of underwater illumination." The remaining six species in Babine Lake undergo little if any vertical migration.

Narver (1970) also describes the diel vertical migrations of juvenile sockeye salmon (*Oncorhynchus nerka*) in Babine Lake. The juvenile sockeye spend daylight hours at depths of 25–50 m, rising towards the surface (0–10 m) for brief periods at dawn and dusk, and spend the night at an intermediate level ($\approx$ 13 m); also see Levy (1987). Sockeye salmon are planktivores, the most common food item in this lake being *Daphnia longispina*, followed by *B. coregoni* and *H. septentrionalis*. Zaret (1980) hypothesizes that predation pressure is responsible for the observed migration patterns of zooplankton in Babine Lake: "... the coincidence of these predator and prey migration patterns probably relates to selective predation by size-dependent predators" (Zaret 1980, p. 77). We will discuss Zaret's hypotheses in detail in Section 5.3.

The vertical migrations of the copepod *Cyclops abyssorum* in alpine lakes of the Tatra mountains in Poland have been described by Gliwicz (1986a). These copepods exhibit a wide range of migratory patterns. In lakes containing no predatory fish the copepods do not migrate vertically; in lakes that have been stocked with planktivorous fish for several decades they undertake short-range migrations; in other lakes where such fish have been present for centuries they undertake long-range vertical migrations. Gliwicz concludes that these observations support the hypothesis that vertical migration of zooplankton "is selected for as a means of evading fish predators" (Gliwicz 1986a, p. 746).

The migration patterns of a copepod, *Pseudocalanus* sp., in Puget Sound, Washington, are discussed by Ohman et al. (1983). At certain times of the year *Pseudocalanus* performs reverse diel

Table 5.1
Life history parameters of *Daphnia pulex* (Lynch 1980)

| | | |
|---|---|---|
| average size at birth | 0.69 | mm |
| maximum size | 3.50 | mm |
| average size at first reproduction | 1.40 | mm |
| age at first reproduction | 6.5 | d |
| average life span | 66.0 | d |
| maximum clutch size | 82 | eggs |
| egg volume | $7.4 \times 10^{-2}$ | $mm^3$ |
| proportion of food used for growth | 4 | % |
| number of adult instars | 16 | |

vertical migrations, which Ohman et al. hypothesize to be a response to predation pressure by invertebrate planktivores. These authors also point out that the invertebrate predators are themselves prey to planktivorous species of fish, which elect them in favor of *Pseudocalanus*. They hypothesize that the invertebrates therefore perform normal vertical migrations, to which *Pseudocalanus* responds by performing reverse vertical migrations.

These examples (only a small sample of many cases examined in the literature) strongly suggest that predation at several trophic levels regulates the migratory behavior of zooplankton, and thus ultimately determines the structure of aquatic communities (Zaret 1980). We will now analyze these observations using dynamic behavioral modeling.

### 5.1.1 Cladocerans

The life history strategies of zooplankton consist of two main types. The first, exemplified by the cladocerans, involves continuous parthenogenetic reproduction throughout the summer season (in temperate waters), which may be followed by a final phase of sexual reproduction (Lynch 1980—see our Table 5.1). The second type, exemplified by many species of copepods, involves a small number of episodes of sexual reproduction. In this section we consider dynamic models of vertical migration by zooplankton for two extreme strategies of reproduction: (i) a cladoceran model, with continuous reproduction, and (ii) a copepod model, with a single reproductive episode.

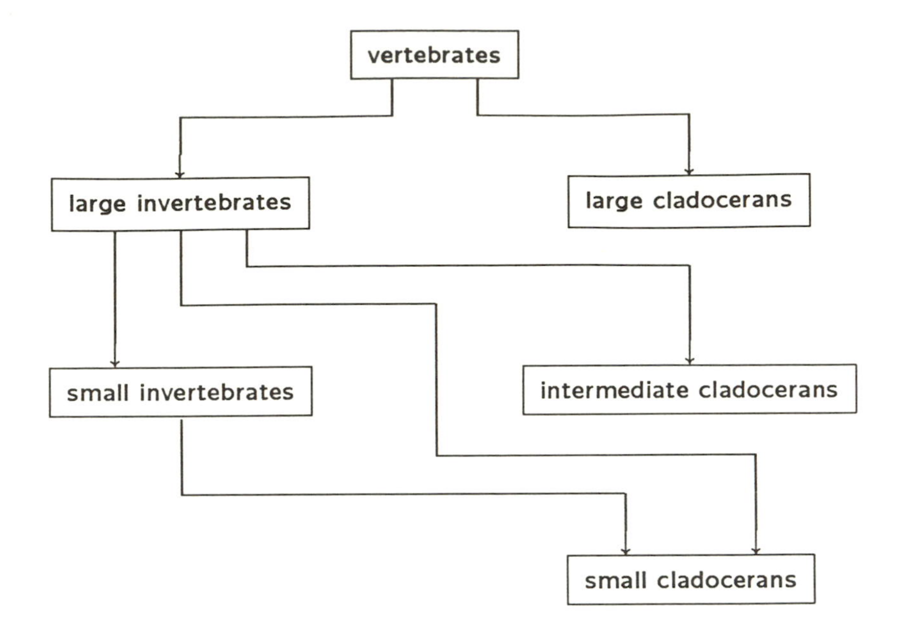


Figure 5.2 Predation patterns in a freshwater community: If present, vertebrate predators (mostly fish) elect large cladocerans, and may also eat large invertebrates, which in turn eat smaller invertebrates and cladocerans. Lakes containing vertebrate predators often contain few large cladocerans or large invertebrates, whereas in lakes devoid of vertebrate predators small cladocerans may be heavily preyed upon by invertebrate species (Lynch 1980).

Cladocerans are small (0.2–5.0 mm) crustaceans. Clutches of eggs are carried in a brood chamber, and released as live young after hatching (Lynch 1980). Predation by both vertebrate and invertebrate predators is often severe. Typical predation patterns are described in Figure 5.2.

The net daily energy intake (energetic value of food consumed, less daily metabolic cost) of a cladoceran generally depends on its body size. The body size that maximizes net daily energy intake is called the "optimal foraging size" by Lynch (1980), who observes that in cases where predation risk is size-dependent, optimal foraging size will not in general be equal to the optimal body size in terms of expected reproductive success. Thus cladoceran populations which are subject to predation by vertebrate predators should evolve body sizes less than the optimal foraging size, while cladocerans subject to heavier predation risks at small size should grow larger than the optimal foraging size. Data in support of this prediction are given by Lynch (1980, Figure 7).

To model the effect of vertical migration strategy on the lifetime fitness (reproductive success) of cladocerans, we consider only the adult, reproductive phase, and make the following simplifying assumptions:

(1) No growth occurs during the adult phase.
(2) Reproduction per period is proportional to food intake less metabolic cost.
(3) If migration takes place, it involves two habitats: $H_1$, a near-surface habitat containing food (phytoplankton) and predators; and $H_2$, a deep water habitat devoid of both food and predators. The predators in $H_1$ may be active during the day, or during the night, but not both.
(4) The daily metabolic cost is negligible if no migration takes place, and is a fixed constant $\alpha > 0$ when vertical migration is undertaken.
(5) The terminal fitness of a surviving cladoceran $\phi_T$ is a fixed constant, representing final sexual reproductive output.

These assumptions are adopted in order to study migration strategy in isolation from other aspects of cladoceran life history. Several extensions which would make the model more complex and realistic are readily conceived. For example, growth of adults could be included, possibly allowing for an optimal allocation of net food intake between growth and reproduction.

With the above assumptions, the lifetime fitness function of a cladoceran can be written as

$$F(t,T) = \max E\{\textstyle\sum_{j=t}^{T-1} R(j) + \phi_T \mid \text{cladoceran is alive at the beginning of period } t\} \tag{5.1}$$

where $R(j)$ denotes reproductive output in period $j$. Because growth is not considered, the model requires no explicit state variable $X(t)$; the state of a cladoceran in period $t$ is simply that it is either alive or dead. The maximization in Eq. (5.1) is taken with respect to migratory behavior, i.e. whether to migrate or not, and if so, how much of the 24-hour daily period $t$ to spend in

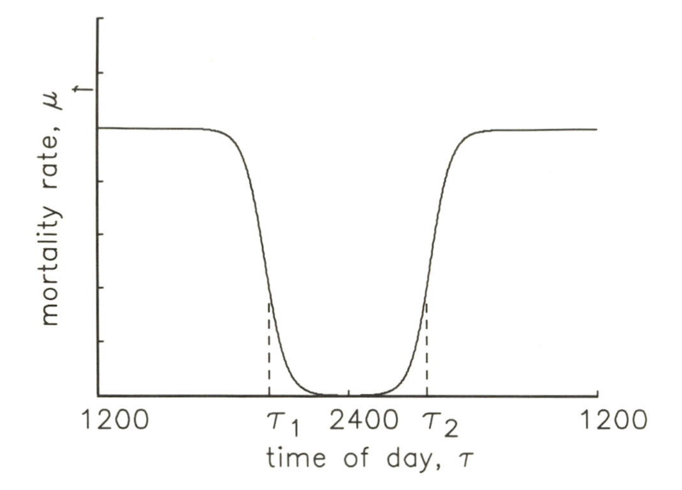


Figure 5.3 Instantaneous mortality risk $\mu(\tau)$ for a zooplankter subject to predation by visually searching planktivores. The interval $[\tau_1, \tau_2]$ minimizes total predation risk for a given daily period of length $s = \tau_2 - \tau_1$ spent in $H_1$.

the surface habitat $H_1$. Reproductive output and mortality risk both depend on this migration pattern.

Let $\mu(\tau)$ denote instantaneous mortality risk in $H_1$, so that

$$\mu(\tau)\,d\tau = \Pr(\text{cladoceran is killed by predator in } H_1 \text{ between times } \tau \text{ and } \tau + d\tau). \tag{5.2}$$

For the case of daytime predation by visually searching planktivores, $\mu(\tau)$ will be primarily determined by light intensity in $H_1$; the graph of $\mu(\tau)$ is qualitatively shown in Figure 5.3, which shows the case in which predation risk is highest during daylight hours.

If the cladoceran spends the time interval from $\tau_1$ to $\tau_1 + s$ in habitat $H_1$, it will be exposed to a total daily mortality risk $m$ given by

$$m = 1 - \exp\left(-\int_{\tau_1}^{\tau_1+s} \mu(\tau)\,d\tau\right). \tag{5.3}$$

Consider first the problem of minimizing this mortality risk, for a given total amount of time $s$ spent in $H_1$. From Eq. (5.3), this is equivalent to minimizing the integral $\int_{\tau_1}^{\tau_1+s} \mu(\tau)\,d\tau$ by an appropriate choice of $\tau_1$. In other words, the area under the curve

$\mu = \mu(\tau)$ between $\tau_1$ and $\tau_1 + s$ must be minimized, with $s$ preassigned. This area will be minimized provided that*

$$\mu(\tau_1) = \mu(\tau_1 + s). \tag{5.4}$$

Assume, for example, that the instantaneous mortality risk $\mu(\tau)$ is directly related to light intensity in $H_1$. Then Eq. (5.4) leads to the prediction that vertical migration into and out of $H_1$ will occur at equal levels of light intensity. This prediction agrees with many observations.

In practice, most of the change in light intensity within a given 24-hour period occurs during relatively brief time periods at dawn and dusk. To simplify our model further, let us assume that if the cladoceran does migrate between $H_1$ and $H_2$ it will do so at fixed times at dawn and dusk; seasonal variations in the timing of dawn and dusk are therefore ignored (but could be included in a more detailed model).

Let $\hat{s} = 24\,\text{hr}$ and let $\hat{m} = m(\hat{s})$ denote the mortality risk if the 24-hr day is spent feeding in $H_1$. Also let $s_1$ denote the time spent in $H_1$ per day if migration takes place, and let $m_1$ denote the corresponding mortality risk. Assume that reproductive output $R$ results only if the cladoceran survives the day, and is proportional to food intake minus the metabolic cost of migration; food intake is proportional to the time spent in $H_1$:

$$R = \begin{cases} \gamma\hat{s} & \text{if } \hat{s} = 24\,\text{hr are spent in } H_1 \\ \gamma(s_1 - \alpha) & \text{if } s_1 < 24 \text{ hr are spent in } H_1 \end{cases} \tag{5.5}$$

where $\gamma$ = a constant, and where metabolic cost $\alpha$ is measured in terms of the feeding time required to make up for the energy expended in (two) migrations.

The dynamic programming equation for our cladoceran model is then

$$F(T,T) = \phi_T \tag{5.6}$$

* To see this, suppose that condition (5.4) is violated, for example with $\mu(\tau_1) < \mu(\tau_1 + s)$. Then decreasing $\tau_1$ slightly will decrease the area under the curve (try sketching this), so that $\tau_1$ is not the minimizing value unless (5.4) holds. An alternative proof based on calculus goes as follows: write $\int_{\tau_1}^{\tau_1+s} \mu(\tau)\,d\tau = F(\tau_1 + s) - F(\tau_1)$, where $F'(\tau) = \mu(\tau)$. Minimizing this expression requires that $\frac{d}{d\tau_1}[F(\tau_1 + s) - F(\tau_1)] = 0$, or $\mu(\tau_1 + s) - \mu(\tau_1) = 0$, as before.

$$F(t,T) = \max[(1-\hat{m})(\gamma\hat{s} + F(t+1,T)),$$
$$(1-m_1)(\gamma(s_1-\alpha) + F(t+1,T))] \quad \text{for } t < T \quad (5.7)$$

where the first term corresponds to no migration between $H_1$ and $H_2$ and the second term corresponds to migration. We have:

$$\hat{m} > m_1 \qquad \text{and} \qquad \hat{s} > s_1.$$

The dynamic programming equation (5.7) is so simple that a graphical solution is possible. First we further simplify the notation. Let $p_1 = 1 - m_1$, $\hat{p} = 1 - \hat{m}$, and $f_1 = \gamma(s_1 - \alpha)$, $\hat{f} = \gamma\hat{s}$. Also let $F_n = F(T-n,T)$, so that $F_n$ represents lifetime fitness with $n$ days to go. Equation (5.7) becomes now

$$F_n = \max[p_1(f_1 + F_{n-1}), \hat{p}(\hat{f} + F_{n-1})] \qquad (5.8)$$

where

$$\hat{p} < p_1 \qquad \text{and} \qquad \hat{f} > f_1 \qquad (5.9)$$

i.e. remaining in $H_1$ for the entire day is both riskier and more productive than migration to $H_2$ for part of the day. Equation (5.8) can be expressed in the simple form

$$F_n = Q(F_{n-1})$$

The sequence $\{F_n\}$ generated by such a relation can be constructed graphically as shown in Figure 5.4.

For Eq. (5.8) we have

$$Q(F) = \max[p_1(f_1 + F), \hat{p}(\hat{f} + F)]. \qquad (5.10)$$

Two examples are shown in Figure 5.5, where the heavy lines represent the graph of the function $Q(F)$, i.e. the maximum of the two linear expressions in $F$. The value of $F$ at which the maximum in Eq. (5.10) switches is labeled $\hat{F}$ in Figure 5.5, and is given by $\hat{F} = (\hat{p}\hat{f} - p_1 f_1)/(p_1 - \hat{p})$. If $F_{n-1} < \hat{F}$ the optimal strategy with $n$ days to go is not to migrate, but to spend the whole day feeding in $H_1$; for $F_{n-1} > \hat{F}$ migration is the optimal strategy.

Now consider the graphical solution of Eq. (5.8). For $n = 0$ we have $F_0 = F(T,T) = \phi_T$ = expected terminal (sexual) reproduction. Suppose $\phi_T < \hat{F}$; then nonmigration is the optimal strategy

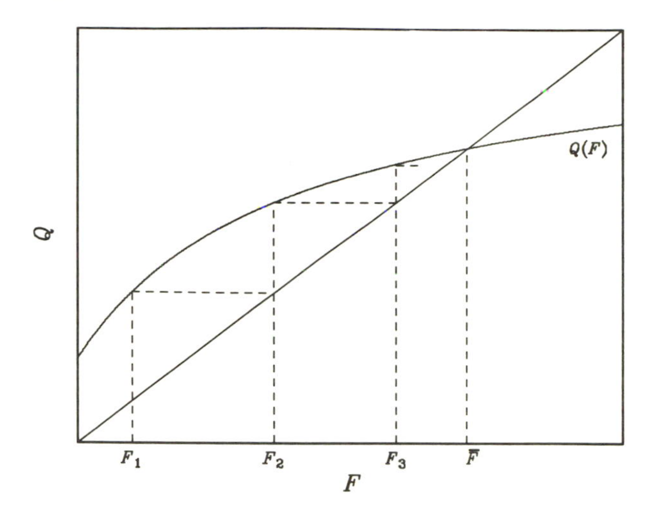


Figure 5.4 The evolution of the relation $F_n = Q(F_{n-1})$: $F_n$ is obtained from $F_{n-1}$ by the construction: move from $F = F_{n-1}$ up to the curve $Q(F)$, then over to the 45° transfer line, obtaining $F_n$. In the case shown, $F_n$ increases monotonically towards the equilibrium $\bar{F}$ defined by $\bar{F} = Q(\bar{F})$.

for $n = 1$ day to go. In the case shown in Figure 5.5(a), we have $F_1 = Q(F_0) > \hat{F}$, so that migration becomes the optimal strategy for $n = 2$. Consequently migration is also optimal for all $n \geq 2$. As $n$ increases, lifetime fitness $F_n$ increases towards the limit $\bar{F}$.

(By the way, an interesting possibility arises if $F_0 = \phi_T > \bar{F}$, for in this case lifetime fitness $F_n$ *decreases* as $n$ increases. This result may seem surprising, but it is easily explained. When terminal reproduction $\phi_T$ is large, the cladoceran may *decrease* its expected lifetime fitness by exposing itself to predators, since the expected loss in sexual reproduction ($\phi_T$ times the probability of predation) is greater than the possible gain from parthenogenetic reproduction. In this situation, a third strategy of remaining permanently in the deep-water habitat $H_2$ would in fact be a superior strategy. Some copepod species are known to employ such a strategy seasonally, and explicit allowance for it will be made in the copepod model discussed later.)

Figure 5.5(b) shows a case in which nonmigration remains optimal throughout the cladoceran's adult phase (assuming that

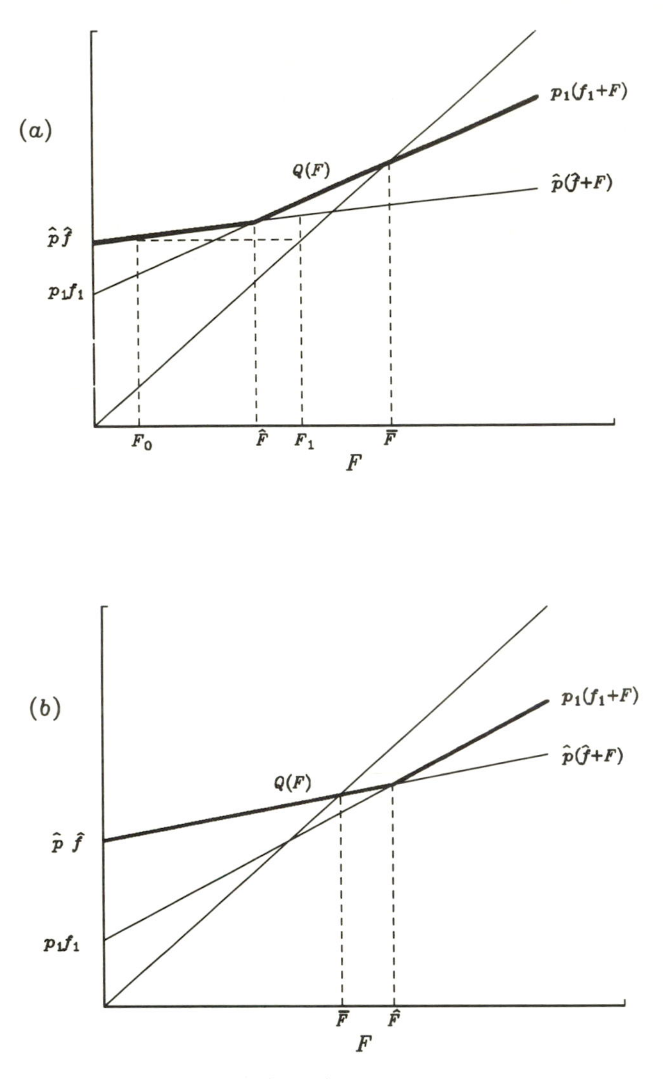


Figure 5.5 Graph of Eq. (5.10), and solution of the dynamic programming equation (5.8); $\hat{F}$ denotes the value of $F$ at which the optimal migration strategy switches from nonmigration to migration, and $\bar{F}$ denotes the steady-state value of $F$.

$\phi_T < \bar{F}$; if $\phi_T > \bar{F}$ then the strategy of staying entirely in $H_2$ becomes optimal, as noted above). This case arises if food intake $f_1$ is sufficiently large relative to the mortality risk $m_1$ in $H_1$. Stich and Lampert (1981) describe the vertical migrations of two species of *Daphnia*, *D. hyalina* and *D. galeata* in Lake Constance, Germany. The former species performs diel vertical migrations, whereas the latter does not. The authors hypothesize that the different strategies can be accounted for by differentials in fecundity and predation pressure between the two species of *Daphnia*. Our model supports this hypothesis.

The insight obtained from this simple model of vertical migration of cladocerans is that optimal migration strategy depends not only on the tradeoff between current food intake and predation risk, but also on the amount of expected future reproduction that is put at risk when entering the dangerous surface habitat. When expected future reproduction $F(t+1, T)$ is high, the optimal strategy may be not to enter the surface habitat $H_1$ at all. Conversely, when expected future reproduction is low, it may be optimal to spend the entire 24-hour day in $H_1$. Intermediate cases give rise to optimal migrations between $H_1$ and $H_2$.

This model can be applied to zooplankton species other than cladocerans. For example, as remarked earlier, Ohman et al. (1983) discuss the reverse diel vertical migrations of *Pseudocalanus*, a continuously reproducing copepod, in Puget Sound, Washington. The following parameter values apply to this species:

| | Migrants | Nonmigrants |
|---|---|---|
| survival probability (1 day) | 0.9768 | 0.9541 |
| fecundity (eggs/day) | 2.029 | 2.209 |

Assuming that $\alpha = 0$ (metabolic costs are usually negligible for copepods; Vlymen 1970), we see that these values imply $\hat{F} = 5.54$ eggs. Assuming zero terminal reproduction, $\phi_T = 0$, and using Eq. (5.8), we obtain the results shown in Table 5.2.

The optimal strategy for *Pseudocalanus* is to migrate, except for the final three days. (Ohman et al. (1983) report that *Pseudocalanus* stops migrating after August, but this is attributed to a reduction of predation pressure at that time.) Table 5.2 also

Table 5.2
Optimal migration strategy, maximum expected fitness, total reproduction, and nonmigrant's fitness, for the cladoceran model applied to *Pseudocalanus* data

| Days to Go, $n$ | Optimal Strategy* | Maximum Fitness, $F_n$ | Nonmigrant's Fitness, $G_n$ |
|---|---|---|---|
| 1 | $X$ | 2.11 | 2.11 |
| 2 | $X$ | 4.12 | 4.12 |
| 3 | $M$ | 6.04 | 6.04 |
| 4 | $M$ | 7.88 | 7.87 |
| 5 | $M$ | 9.68 | 9.61 |
| 10 | $M$ | 18.07 | 17.22 |
| 50 | $M$ | 59.09 | 41.54 |
| 100 | $M$ | 77.28 | 45.50 |

* $X$ means do not migrate, $M$ means migrate.

compares the expected lifetime reproduction of a migratory and a nonmigratory copepod: over a 100-day horizon, migration increases expected reproduction by 70%. Ohman et al. (1983) used a life-history model to show that migratory strategy was superior to nonmigration for *Pseudocalanus*, but their model did not allow for the switch in strategy that the dynamic model does.

### 5.1.2 Copepods

We now consider a model of semelparous reproduction by a zooplankter such as a copepod. We shall not address the question of comparing the fitness of iteroparous and semelparous life history strategies (cf. Murphy 1968, Tuljapurkar 1982).

As state variable we take $X(t)$ = weight of reproductive tissue (or potential reproductive tissue). We consider three possible migration strategies $S_i$ between the near surface habitat $H_1$ containing food, and the deep water habitat $H_2$ with no food. The strategies are:

$S_1$: spend the entire 24-hour day in $H_1$
$S_2$: spend the entire day in $H_2$
$S_3$: migrate between $H_1$ and $H_2$

We do not attempt to model the optimal timing of migration (see Section 5.3 below), but simply assume that the zooplankter migrates at fixed times to avoid predators in the surface habitat $H_1$. Strategy $S_i$ involves a daily metabolic cost $\alpha_i$ (which decreases potential reproductive tissue), and incurs a risk $\beta_i$ of predation, where the $\beta_i$ are ordered as follows:

$$\beta_1 > \beta_3 > \beta_2 = 0. \qquad (5.11)$$

Food intake resulting from strategy $S_i$ is a random variable $Z_i$ with distribution

$$\Pr(Z_i = Y_{ij}) = \lambda_{ij} \qquad j = 1, 2, \ldots, J_i. \qquad (5.12)$$

Since $H_1$ contains food while $H_2$ does not, we have

$$\bar{Z}_1 > \bar{Z}_3 > \bar{Z}_2 = 0 \qquad (5.13)$$

(where $\bar{Z}_i$ is the mean of $Z_i$). If $S_i$ is chosen, then the dynamics of the state variable $X(t)$ are given by

$$X(t+1) = X(t) - \alpha_i + \gamma Z_i \qquad (5.14)$$

where $\gamma$ is a food-to-reproductive-tissue conversion factor. Also, $X(t)$ is constrained by

$$0 \leq X(t) \leq C. \qquad (5.15)$$

A copepod does not necessarily starve when reproductive tissue falls to zero, so we will *not* assume that $X(t) = 0$ corresponds to death. (The metabolic cost $\alpha_i$ is assumed to be covered by some other variable, such as fat reserves, if $X(t) = 0$.) Thus the only source of mortality in this model is predation.

Finally, we assume that reproductive output is proportional to terminal reproductive tissue $X(T)$:

$$\phi(X(T)) = kX(T) \qquad k > 0. \qquad (5.16)$$

Lifetime fitness is then defined by $F(x, t, T) = \max E\{\phi(X(T)) \mid X(t) = x\}$. The dynamic programming equation is:

$$F(x, t, T) = \max_{i=1,2,3} (1 - \beta_i) \sum_{j=1}^{J_i} \lambda_{ij} F(x'_{ij}, t+1, T) \quad \text{for} \quad x \geq 0 \qquad (5.17)$$

where

$$x'_{ij} = \text{chop}(x - \alpha_i + Z_{ij}; 0, C). \tag{5.18}$$

Some qualitative insights can be obtained from the above equation directly, without any numerical computation. If $X(t)$ is large, especially if $X(t)$ is near $C$, the benefits from additional feeding may not be worth the risk of losing the entire reproductive "asset" $X(t)$. Thus $S_2$ should be the optimal strategy when $X(t)$ is large, whereas $S_1$ or $S_3$ should be optimal for smaller values of $X(t)$. Such changes in migration strategy are discussed briefly by Baker (1978); they are usually attributed to seasonal changes in predation risk, rather than intrinsic factors involving expected future reproduction. Our model could also easily be modified to allow predation risk to depend on the state variable, $\beta_i = \beta_i(x)$. The dynamic programming equation (5.17) then uses $\beta_i(x)$ in place of $\beta_i$. This change would facilitate an analysis of the effects of ontogeny on migratory behavior—see Zaret 1980, Werner and Gilliam 1984.

If forage is highly patchy, so that the daily increment to $X(t)$ is a random variable with high variance, and if food is a limiting factor in the growth of reproductive tissue, then among members of a given population one would expect to observe considerable variation in size. Our dynamic state variable model predicts that large and small copepods of the same population might behave differently; large copepods should be more sensitive to predation risk, and hence less likely to migrate, than small copepods. Such behavioral differences have been observed in some situations (e.g. Narver 1970). According to our model, this variation would occur even if large and small copepods were subject to equal predation risks.

The models considered in this section have deliberately been kept very simple, since our main purpose is to indicate how the dynamic approach can be applied to the study of vertical migration of zooplankton. Many additional details could be included (subject to the availability of data), such as the effects of water temperature on growth and development (Kerfoot 1970), or the relationship between body size and predation risk (Zaret 1980). Alternative approaches to the modeling of the tradeoff between food intake and predation risk for aquatic organisms include traditional life history strategy models (e.g. Ohman et al. 1983), and optimal control theory (e.g. Werner and Gilliam 1984). The approach followed here, while closely related to both these methods, seems to offer the dual advantages of simplicity and flexibility.

## 5.2 Diel Migrations of Planktivores

Figure 5.6 shows the diel pattern in the vertical distribution of juvenile sockeye salmon (*Oncorhynchus nerka*), on a typical summer day in Babine Lake, British Columbia. Juvenile salmon spend daylight hours in deep water (17–36 m depth), and ascend towards the surface at dusk. For about 30 min after twilight (2125–2145 hr) the fish feed actively on zooplankton at or near the surface, and then disperse, remaining within the top 15 m throughout the night. At the first light of dawn at least some of the fish again approach the surface for about 30 min, before descending to their daytime levels (Narver 1970, p. 289). This migration pattern is consistent from year to year; similar migrations of young sockeye have been observed in many other lakes (Levy 1987).

The diel foraging and migratory behavior of golden shiners (*Notemigonus crysoleucas*) in Warner Lake, Michigan, has been described by Hall et al. (1979); see Figure 5.7. Golden shiners spend daylight hours in the littoral zone, moving offshore 10–25 min after sunset to spend the next 45–100 min feeding on zooplankton (mainly *Daphnia*); the pattern is reversed at dawn. Hall et al. (1979, p. 1037) quote five other examples of similar movements by planktivorous fish; see also Wurtsbaugh and Li (1985).

What is the adaptive significance of these patterns of movement? Narver (1970 p. 313) says: "Presumably the diel vertical movements of young sockeye salmon in the lake are a manifestation of natural selection, but exactly what the selective pressures are (or were) is unknown." In discussing the possible effects of predation on salmon behavior, Narver says: "It seems that a fish that spends most of its life in twilight would be less likely to be preyed upon than would the same fish found in higher intensity of illumination. However, there seem to be few predators of underyearling sockeye in the limnetic zone." Night purse seine sets yielded about one rainbow trout (*Salmo gairdneri*) or lake trout (*Salvelinus namaycush*) per set, which Narver considers too few predators to affect sockeye behavior. However, if such a trout kills several sockeye per day, the predation rate over the sockeye's first year in the lake may be substantial. In some lakes juvenile sockeye suffer annual mortality rates up to 90% (Foerster 1968), presumably largely from predation. Hall et al. (1979) report that golden shiners are the preferred prey of large piscivores, such as largemouth bass (*Micropterus salmoides*), which are common in Warner Lake.

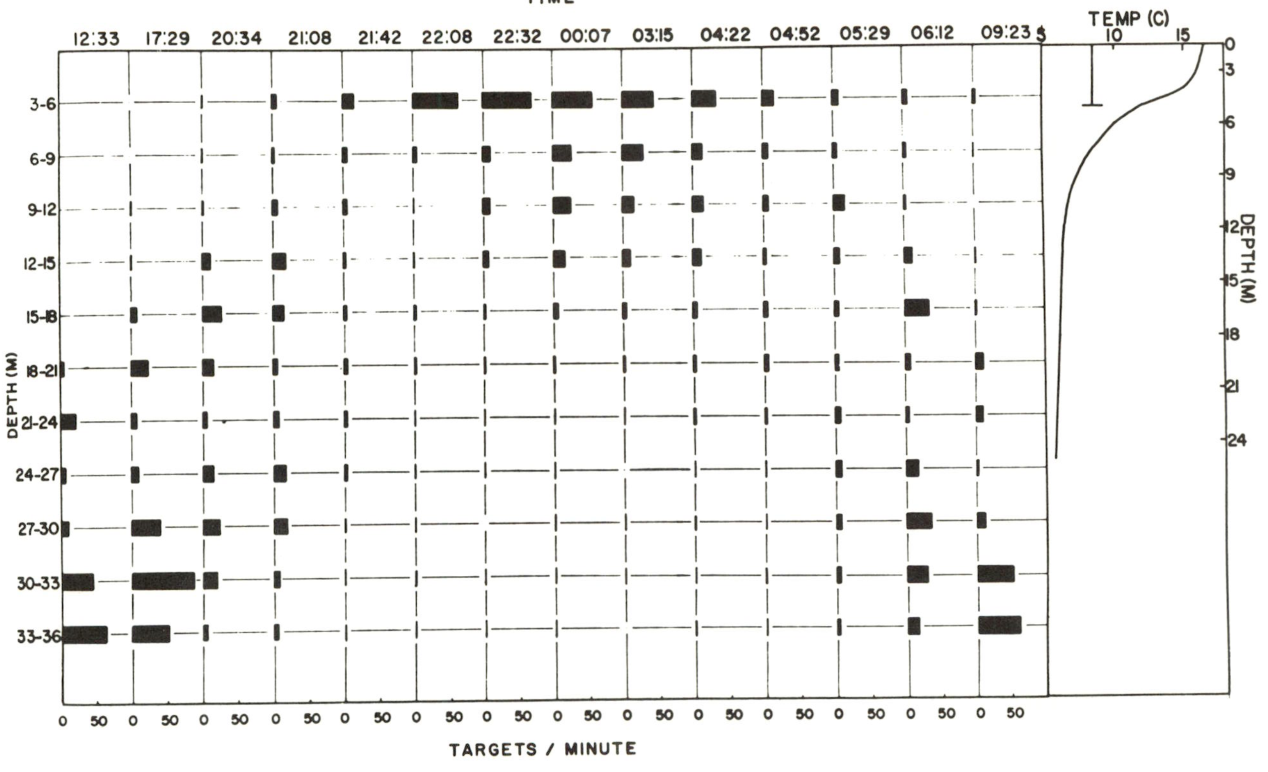

Figure 5.6 Diel vertical movements of young sockeye salmon in Babine Lake, August 5–6, 1967. Bar lengths are proportional to number of echosounder targets per minute of towing. Sunset at 2055 hr and sunrise at 0515 hr; civil twilight at 2140 and 0430 hr PDST (Narver 1970).

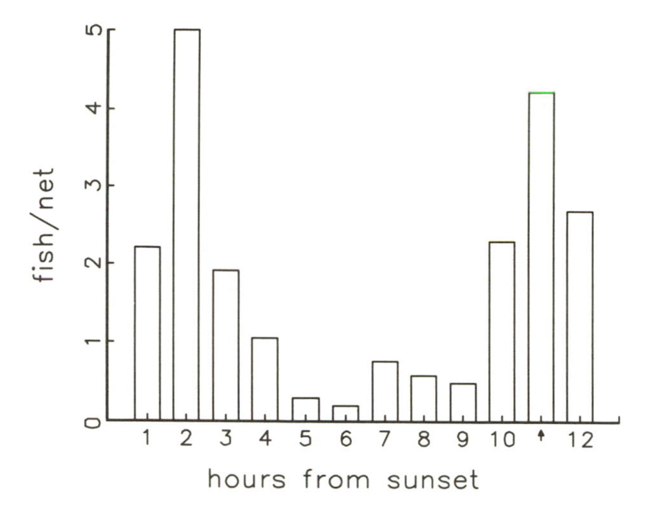


Figure 5.7 Average number of golden shiners caught per net per hour standardized to time of sunset (zero hour) in Warner Lake (Hall et al. 1979). The arrow indicates average time of sunrise.

We consider the hypothesis that the diel migrations of species such as juvenile sockeye and golden shiners are an adaptive response to the tradeoff between food intake and predation risk. We will construct a dynamic model to see whether this hypothesis is capable of predicting the observed timing of these migrations.

Planktivores such as juvenile sockeye and golden shiners locate prey primarily by visual search. Similarly, the larger species that prey upon these planktivores are also visual searchers. Consequently both the feeding rate of the planktivore, and its risk of predation should depend critically on light intensity. Feeding rate and predation risk will be high in illuminated waters containing zooplankton, and both will be low under conditions of darkness. But why should surface feeding just at twilight be advantageous?

### 5.2.1 A Model of Aquatic Predation

In order to address this question, we first require an explicit model of the relationship between visual attack range and aquatic feeding rates. Consider a fish searching for prey, swimming with cruise speed $v$. Assume, for simplicity, a circular field of vision, with radius $r$ (depending on light intensity); see Figure 5.8. Prey are

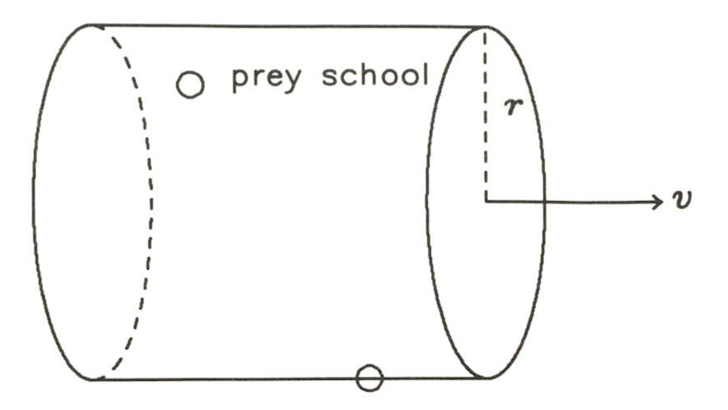


Figure 5.8 The cylindrical search volume of a visual aquatic predator (after Eggers 1976, Clark and Levy 1988).

assumed to occur in spherical schools of radius $r_0$, each containing $N$ prey. The density of prey schools is denoted by $\rho$. A school is sighted if any portion of it passes within the search range of the predator. Once sighted, a school is attacked, and an average of $\epsilon$ prey are consumed, requiring a total attack and handling time of $\tau$.

In a long search time $T$, an effective volume $V = \pi(r + r_0)^2 vT$ will be searched, and $V\rho$ schools will be sighted. Hence the long-term average feeding rate $f(r)$ will be equal to

$$\begin{aligned} f(r) &= \frac{\text{number of prey eaten}}{\text{total search time} + \text{total handling time}} \\ &= \frac{\epsilon V \rho}{T + \tau V \rho} \\ &= \frac{\epsilon/\tau}{T/(V\rho\tau) + 1} = \frac{\epsilon/\tau}{1 + T/(\pi(r + r_0)^2 vT\rho\tau)} \\ &= \frac{\epsilon/\tau}{1 + Q^2/(r + r_0)^2} \end{aligned} \tag{5.19}$$

where

$$Q^2 = \frac{1}{\pi v \tau \rho}. \tag{5.20}$$

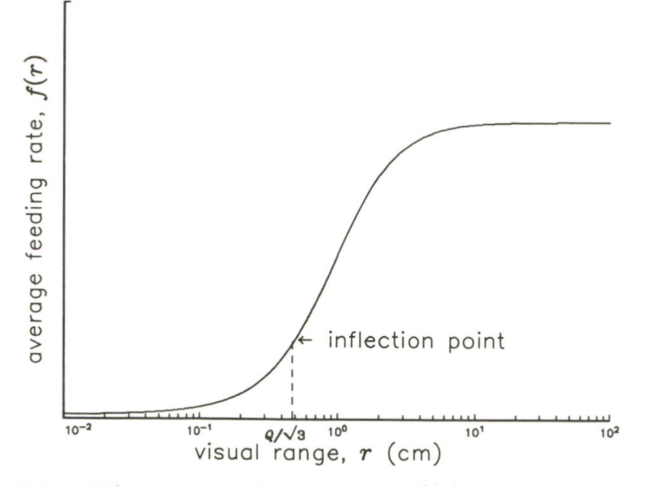


Figure 5.9 The average feeding rate $f(r)$, as a function of visual range, $r$. See Eq. (5.19).

Note that the function $f(r)$ depends on only three parameter combinations: $\epsilon/\tau$ = average number of prey consumed per encounter, $r_0$ = radius of prey schools, and $Q$. The graph of $f(r)$ is as shown in Figure 5.9; one can show by calculus that $f(r)$ is monotone increasing, with asymptotic limit $f_\infty = \epsilon/\tau$ as $r \to +\infty$, and also that the graph has an inflection point at $r + r_0 = Q/\sqrt{3}$. (Note that $f_\infty$ can be interpreted as the feeding rate when search time is negligible.) Predation curves having this shape have been called "Type III" functional responses by Holling (1965).

Parameter estimates, using golden shiners as an example, can be obtained from the data given by Hall et al. (1979). Cruise speed is often quoted to be about one body length per second: $v = 13.5 \times 10^{-2}\,\text{ms}^{-1}$. Hall et al. state that shiners eat about 2 *Daphnia* per second under well-illuminated laboratory conditions, so that $\tau \approx 0.5\,\text{s}$ (see also Werner 1977). The density of *Daphnia* in offshore waters in Warner Lake is about 20–40 per liter, so that $\rho \approx 3 \times 10^4/\text{m}^3$. Copepods do not form schools, so that $r_0$ ("radius" of one copepod) is about 0.5 mm, which is negligible. The curve plotted in Figure 5.9 uses these parameter values, and has its inflection point at $7.2 \times 10^{-3}\,\text{m}$. Thus reduced light intensity begins to affect the feeding rate of such a planktivore when its visual attack range has been reduced to approximately 1 cm.

Next consider the case of the shiners' predators, primarily largemouth bass. Hall et al. (1979) estimated a population of 2500 golden shiners in Warner Lake, which has a surface area of 46 ha. The littoral zone extends about 100 m from shore, giving an approximate area of the offshore waters equal to 25 ha. The shiners were largely confined to the top meter of the water column, so that average density was about $\rho = 10^{-2}/\text{m}^3$. Golden shiners do not school while feeding. Details of predation by largemouth bass are not given, so we will assume $v = .3\,\text{ms}^{-1}$ and $\tau = 5\,\text{s}$. The feeding curve $f(r)$ for largemouth bass preying on golden shiners thus has its inflection point at $r = 2.5\,\text{m}$ (since $r_0 = .135\,\text{m}$, the body length of a shiner).

We assume that the risk of predation $\mu$ to an individual shiner is proportional to the feeding rate of largemouth bass (and similar predators). The potential predation rate of largemouth bass on golden shiners thus begins to decline severely when the visual attack range has fallen to about 2.5 m; at 1 m range, the potential predation rate has declined to 5% of its maximum daylight value. When the shiners' visual range is 1 m, on the other hand, they still maintain well over 99.9% of their maximum feeding efficiency. This phenomenon is easily understood intuitively: *Daphnia* are sufficiently dense ($\sim$ 30 per liter) that shiners experience little difficulty in locating *Daphnia* until the visual detection range has fallen to a value of about 1 cm, whereas the relatively low density of shiners ($\sim$ one per $100\,\text{m}^3$) means that the piscivores' search efficiency declines sharply when $r < 2.5\,\text{m}$.

We believe that this very simple scaling argument, comparing the relative positions of the two feeding curves $f(r)$ for the two predator-prey systems, provides considerable insight into the adaptive advantages to be gained by planktivores with piscivorous predators, restricting their surface feeding activities to the twilight zone. (Clark and Levy 1988 refer to this zone as the "antipredation window.") It remains to relate the visual attack range, or reaction distance $r$, to surface illumination. Blaxter (1980) points out that most fish stop feeding at $10^{-1}$ to $10^{-2}$ lux, which he refers to as the "visual feeding threshold." Within this range, reaction distance appears to be a smoothly increasing function of illumination (Vinyard and O'Brien 1976, Henderson and Northcote 1985)—see Figure 5.10. Since illumination in near-surface waters is itself a function of the time of day $t$, we can therefore obtain an empirical estimate for the function $r = r(t)$ relating reaction distance $r$ to

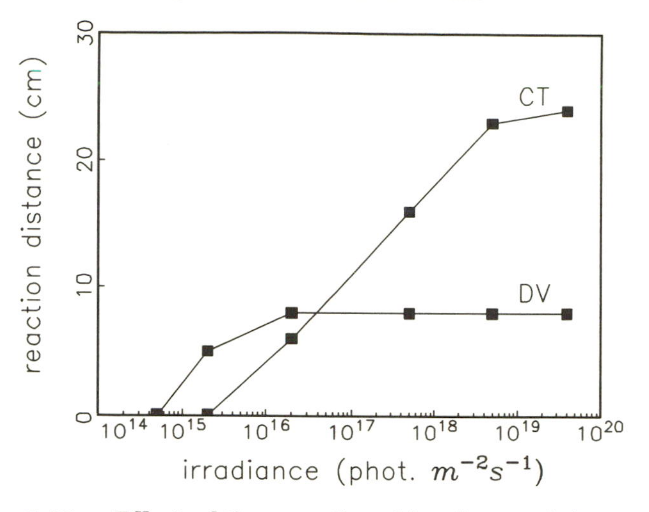


**Figure 5.10** Effect of the quantity of irradiance of the mean reaction distance of cutthroat trout (CT) and Dolly Varden (DV) to natural prey, *Diaptomus kenai* (from Henderson and Northcote 1985).

the time of day. One question remains: How do the visual ranges $r$ of the two predators compare at different low levels of light intensity? Our observations of sockeye in Babine Lake suggest that copepods are still quite visible (to the human eye at least), but that the salmon are almost invisible, at twilight. We guess that in fact the visual ranges for the two predators are quite similar, but this should be tested experimentally. Since the horizontal separation of the two predation curves (in terms of their $Q$ values) exceeds two orders of magnitude, the existence of a twilight antipredation window is assured unless there is a compensating two orders of magnitude difference in the visual ranges of the two predators. Proceeding on the basis of the assumption that the two visual ranges are roughly equal, we next develop a dynamic model to predict the vertical migration strategy of planktivores (see Clark and Levy 1988).

### 5.2.2 A Dynamic Model of Diel Migrations

We now simplify the habitat structure of the (model) lake, by assuming only two habitats: $H_1$, a near-surface, open water habitat containing zooplankton as prey, and also containing piscivorous predators on the planktivores; and $H_2$, a deep water or pro-

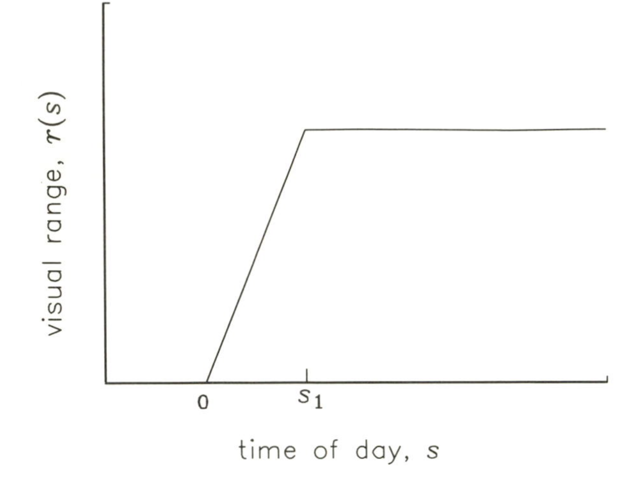


Figure 5.11 Assumed relation between visual range $r(s)$ and time of day, $s$, during the morning twilight period.

tected littoral habitat essentially devoid of prey and piscivores. Feeding rate and mortality risk for planktivores in $H_1$ are denoted by $f(r(s))$ and $\mu(r(s))$, respectively, where $s$ denotes time of day. We assume that the morning and afternoon periods are symmetric, and only consider the morning period, for $s = 0\,\text{hr}$ to $s = 1200\,\text{hr}$. In order to estimate the functions $f(r(s))$ and $\mu(r(s))$ as in Eq. (5.19), we would have to know the light intensity profile $i(s)$ and the reactive distance $r(i)$. These data are not currently available and we will simply assume a linear relationship between visual range (i.e. attack distance) $r$ and time $s$ during this twilight period $s_0 \leq s \leq s_1$ (see Figure 5.11). The functions $f(r)$ and $\mu(r)$ are assumed to have the form of Eq. (5.19), with asymptotic values $f_\infty$ and $\mu_\infty$ which may in general depend upon body weight $x$ of the planktivore. Specifically, following Clark and Levy (1988), we assume that, for the case of juvenile sockeye salmon,

$$f(r;x) = \frac{f_\infty x}{1 + Q_f^2/(r + r_f)^2}$$

$$\mu(r,x) = \frac{\mu_1/(1 + \mu_2 x)}{1 + Q_\mu^2/(r + r_\mu)^2}.$$

Thus feeding rate is proportional to body weight $x$, and predation risk varies inversely with $x$.

Given the functions $f$ and $\mu$, the question of the optimal tradeoff between food intake and risk of predation can now be addressed by means of a dynamic model. The short term "decision" is the interval of time devoted to feeding in habitat $H_1$, per halfday, while the long term objective is lifetime fitness in terms of reproductive success.

Let $\Delta(t)$ denote the feeding interval (in $H_1$) in period $t$ (a halfday), and let $X(t)$ denote body weight. Then total food intake $I$ and mortality risk $M$ over the period $\Delta$ are given by

$$I(x,\Delta) = \int_{\Delta} f(r(s);x)\,ds \tag{5.21}$$

$$M(x,\Delta) = 1 - \exp(-\int_{\Delta} \mu(r(s);x)\,ds). \tag{5.22}$$

Equation (5.22) can be derived by the same method used for the Poisson process in Chapter 1—see Eq. (1.25)ff. (In Eqs. (5.21) and (5.22) we assume for simplicity that body weight $x$ does not change significantly during a given feeding interval.)

Future lifetime fitness (expected reproduction) at the end of the first-year, lake-dwelling phase of sockeye salmon is denoted by $\phi(x)$, where $x = X(T)$. This function could be estimated empirically by tagging fish and subsequently estimating survival and fecundity, relating these to size $x$. Given that such data do not exist at present, we are forced to assume some ad hoc functional form for $\phi(x)$ for the numerical computations. The form assumed is

$$\phi(x) = \begin{cases} 0 & \text{for } x < x_c \\ 1 & \text{for } x \geq x_c \end{cases}$$

where $x_c$ is a critical body weight below which future fitness is zero. As explained in Chapter 2, the particular form assumed for $\phi(x)$ will have little or no effect on the optimal migration pattern except possibly for values of $t$ close to the horizon $T$.

The lifetime fitness function is again defined as

$$F(x,t,T) = \max E\{\phi(X(T)) \mid X(t) = x\}.$$

We then have the dynamic programming equation

$$F(x,T,T) = \phi(x) \tag{5.23}$$

$$F(x,t,T) = \max_{\Delta}(1 - M(x,\Delta))F(x',t+1,T), \qquad t < T \tag{5.24}$$

where

$$x' = x - \alpha + \gamma I(x, \Delta). \tag{5.25}$$

Here $\alpha$ denotes metabolic cost per period ($\alpha$ may also depend on $x$), and $\gamma$ is a food assimilation coefficient.

The following algorithm for the computer solution of Eq. (5.24) is described by Clark and Levy (1988). Suppose that food consumed in period $t$ is equal to $I$; we first treat $I$ as a parameter. Then Eq. (5.24) implies that $\Delta$ must be chosen so that mortality $M(x, \Delta)$ is minimized, subject to the constraint $I(x, \Delta) = I$:

$$\underset{s_1, s_2}{\text{minimize}} \quad (1 - \exp(-\int_{s_1}^{s_2} \mu(r(s); x)\, ds) \tag{5.26}$$

$$\text{subject to} \quad \int_{s_1}^{s_2} f(r(s); x)\, ds = I \tag{5.27}$$

where $\Delta = [s_1, s_2]$.

Equation (5.26) is similar to (5.3), but we must now also consider the constraint (5.27). This problem is easily solved by the method of Lagrange multipliers, as described in most calculus texts. We will outline this method. To start, introduce the Lagrangian expression

$$L = 1 - \exp(-\int_{s_1}^{s_2} \mu(r(s); x)\, ds) + \lambda(\int_{s_1}^{s_2} f(r(s); x)\, ds - I)$$

where $\lambda$ is an unknown multiplier. The first-order necessary conditions for a solution to the constrained minimization problem (5.26), (5.27) are then simply that

$$\frac{\partial L}{\partial s_1} = 0 \qquad \text{and} \qquad \frac{\partial L}{\partial s_2} = 0.$$

These two equations, together with the constraint equation (5.27), can then be solved in principle for the three unknown values $s_1$, $s_2$, and $\lambda$.

Carrying out the differentiations of $L$, we obtain*

$$\exp(-\cdots) \cdot (-\mu(r(s_2); x) + \lambda f(r(s_2); x) = 0$$
$$\exp(-\cdots) \cdot (\mu(r(s_1); x) - \lambda f(r(s_1); x) = 0$$

* Recall from calculus that $\frac{\partial}{\partial s_2} \int_{s_1}^{s_2} G(s)\, ds = G(s_2)$.

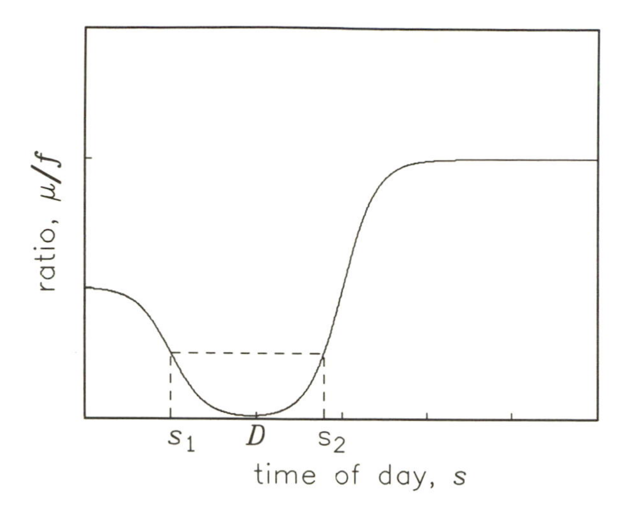


Figure 5.12 The ratio $\mu/f$ of instantaneous mortality risk to food intake, as a function of diel time $s$. The optimal feeding interval $\Delta$ surrounds the time at which $\mu/f$ is minimized.

where $\exp(-\cdots)$ denotes the same exponential expansion as in (5.26). Simplifying, we then conclude that

$$\frac{\mu(r(s_1);x)}{f(r(s_1);x)} = \frac{\mu(r(s_2);x)}{f(r(s_2);x)}. \tag{5.28}$$

Figure 5.12 shows the graph of the ratio $\mu(r(s);x)/f(r(s);x)$ of mortality $\mu$ to food intake $f$. Equation (5.28) implies that the optimal feeding interval, $\Delta = [s_1, s_2]$, is such that the values of $\mu/f$ are the same at $s_1$ and $s_2$ (compare Figure 5.3). Consequently $\Delta$ must contain those points where the ratio $\mu/f$ is minimized, subject to the requirement (5.27) for total food consumption.

The criterion (5.28) as depicted in Figure 5.12 is closely related to the habitat-choice criterion obtained by Werner and Gilliam (1984). By applying the Pontrjagin maximum principle to a continuous-time life history model, Werner and Gilliam show that during the prereproductive phase, the optimal tradeoff between instantaneous predation risk $\mu$ and growth rate $g$ is achieved by minimizing the ratio $\mu/g$. Since we have assumed that growth is proportional to food intake $f$, our criterion (5.28) is closely analogous to that of Werner and Gilliam. (In fact, our criterion remains valid for any monotonic relationship between growth and food intake.)

Let us return to the dynamic programming equation (5.24), which requires a maximization with respect to the feeding interval $\Delta = [s_1, s_2]$. The numerical algorithm that we actually used to find the maximum treats $s_1$ as the only decision variable, and computes $s_2$ from Eq. (5.28), then finds $I(x, \Delta)$ by numerical integration. A search routine is used to determine the optimal value of $s_1$. This problem presents a somewhat greater programming challenge than other models discussed in this book, but the computations are still efficiently carried out on a desktop micro. The results are summarized in Figure 5.13, and described below.

Estimating the parameters and function forms needed to make this model fully operational would be a formidable task, particularly if one wished to include dependence of $f$ and $\mu$ on body size $x$ (and perhaps on calendar date). For the case of juvenile sockeye salmon, Clark and Levy (1988) specify parameters partly on the basis of published data, and partly using informed guesswork. Predation risk parameters are tuned to provide reasonable first-year mortality rates for sockeye. The curves labeled (a) in Figure 5.13 show the growth profile $X(t)$ and survival function $p(t) = \Pr(\text{survive from 0 to } t)$ resulting from the optimal choice of dawn and dusk feeding periods $\Delta(t)$. Initial body weight is assumed to be 0.5 gm, and a terminal weight of 5.0 is assumed necessary for subsequent survival. The optimal feeding interval turns out to be 30 min at dawn and at dusk throughout the assumed juvenile feeding phase of 100 days. Growth proceeds smoothly, approximately exponentially, to the terminal weight, and cumulative mortality risk is 67%.

The curves labeled (b) in Figure 5.13 show growth and survival for a juvenile sockeye that is not adapted to migrations which take advantage of the "antipredation window" at dawn and dusk, but otherwise optimizes fitness. Such a fish grows rapidly (if it survives), but suffers a 98.4% seasonal mortality risk from predation.

As in several of the applications described in this book, the predictions of our dynamic model of vertical migration may ultimately seem anticlimactic. After all, the main prediction, that planktivores should migrate to the surface habitat to feed only for brief periods at dawn and dusk, is already a qualitative prediction of the scaling argument. The dynamic model is capable, at least in principle, of quantifying this prediction.

It has often been our experience that the thought processes

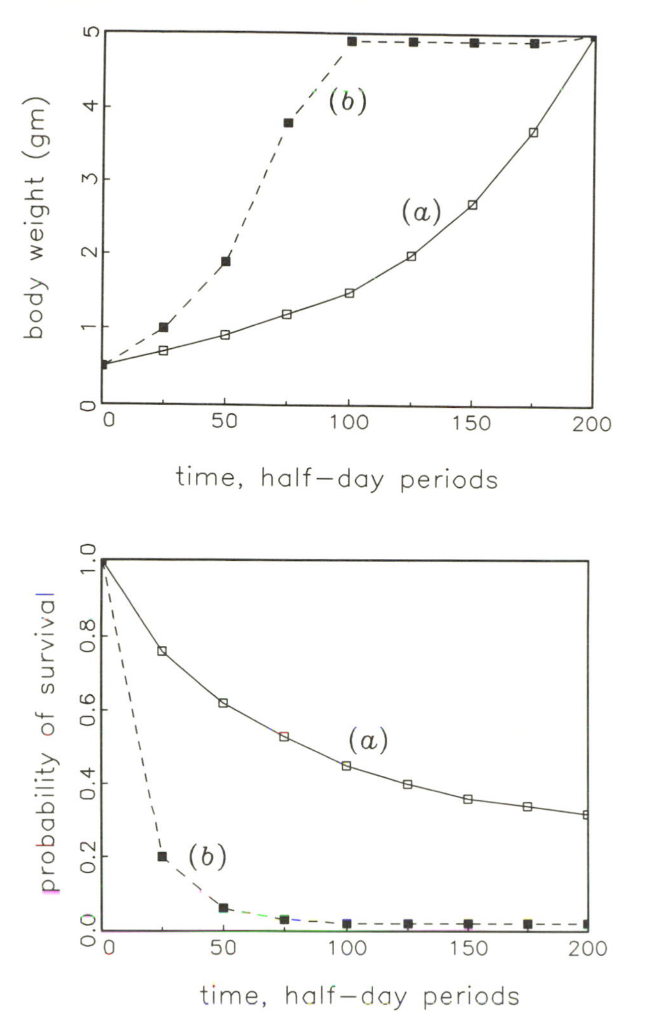


**Figure 5.13** Body weight $x_t$ and probability of survival $p_t$ for (a) optimal crepuscular-feeding planktivore, and (b) noncrepuscular-feeding planktivore (Clark and Levy 1988).

required to construct a dynamic model of some behavioral phenomenon can themselves lead to new insights. In the case of vertical migration, the modeling problem required us to think of some mechanism that would force the model to predict the observed timing of these migrations. The idea of modeling the effects of illumination on aquatic predation rates was then a natural one, and the scaling argument followed easily. We were thus led to formulate a new hypothesis, namely that vertical migration of planktivorous species is a response to the different time profiles of feeding rate and predation risk. As far as we are aware, no other existing hypothesis has the capability of predicting the observed timing of these migrations.

New hypotheses suggest new experiments. For example, it should be possible to perform laboratory measurements of the relation between illumination and feeding rates at different prey densities. To our knowledge such experiments have not yet been performed, but we predict that the results will have the qualitative appearance of the graph of $f(r)$ in Figure 5.9. Moreover, this curve should shift to the left as prey density is increased.

Successful hypotheses should also be capable of explaining phenomena other than those for which they were originally intended. In the next subsection we apply our theory to analyze the migration patterns of zooplankton in lakes containing both planktivores and piscivores.

## 5.3 Predictions of Zooplankton Migrations

Reviewing the observations of Narver (1970) on the vertical migrations of sockeye salmon and their zooplankton prey in Babine Lake, Zaret (1980, p. 77) concludes that "...the coincidence of the predator and prey migration patterns probably relates to selective predation by size-dependent predators." We will now consider this question further. Our analysis indicates that the details of migratory behavior on four trophic levels can be understood on the basis of the salmon's optimal feeding behavior and its effect on the entire community.

The model discussed in Section 5.2.2 simplifies the vertical structure of the lake by considering only two habitats, $H_1$ (near-surface waters) and $H_2$ (deep water). In actuality, the depth structure of a lake is continuously stratified, and light intensity (and spectral composition) varies continuously with both depth and time of day. Thus the "antipredation window" is a spatial as well as a temporal phenomenon. If this window determines the vertical distribution of sockeye, one would predict a 24-hour cycle of sockeye concentration as shown in Figure 5.14. This agrees qualitatively with the observed pattern.

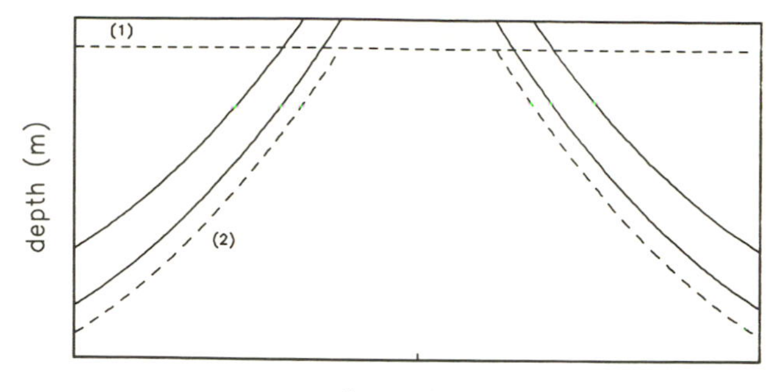


Figure 5.14 Diel changes in the vertical distribution of juvenile sockeye (region between the solid curves), and of optimizing zooplankton (1) and (2)—see text.

Three species of zooplankton in Babine Lake are the primary prey of juvenile sockeye: the cladocerans *Daphnia longispina* (average body size 0.6 mm) and *Bosmina coregoni* (0.4 mm), and the predatory copepod *Heterocope septentrionalis* (3.0 mm). Narver's (1970) observations (Figure 5.1) indicated that *D. longispina* did not undergo vertical migrations, whereas *H. septentrionalis* performed "normal" vertical migrations, spending daylight hours in deep water and rising towards the surface at night. Finally, *B. coregoni* performed "reverse" vertical migrations, descending at night and rising during the day.

Consider a zooplankter subject to predation by juvenile sockeye (Figure 5.14). Assume that the prey of the zooplankter is available only in the near-surface habitat. There are three distinct migration patterns that might optimize zooplankter fitness. First, the zooplankter could spend the entire 24-hour cycle in surface waters, feeding on zooplankton, and facing the risk of predation for brief periods at dawn and dusk (pattern 1 in Figure 5.14). Second, the zooplankter could migrate normally (pattern 2), thereby avoiding its predators completely. At relatively high latitudes, where summer nights are short, pattern 2 would severely restrict the zooplankter's daily feeding time. Finally, a zooplankter could spend daylight hours near the surface, and at some time migrate rapidly downwards through the antipredation window. This strategy (which is not illustrated in Figure 5.14) would be effective against planktivore predation only if the zooplankter were able to descend rapidly through the window.

Which of these three strategies proved to be optimal in any particular situation would depend on several factors, as discussed with regard to the models of Section 5.2. Different strategies could prove optimal at different times of the year. We conjecture that strategy 1 (nonmigration) is optimal for *Daphnia* in Babine Lake, but that strategy 2 (normal vertical migration) is optimal for *Heterocope*, which, because of its large size, would otherwise be the preferred prey of the sockeye. Finally, we agree with Zaret (1980) that the reverse migrations of *Bosmina* are probably a response to the migrations of *Heterocope*, an important predator on *Bosmina*.

The vertical movements of trout species (the salmon's predators) in Babine Lake do not seem to be known, but our theory suggests that the trout would be found where salmon were concentrated, at least when light intensity is sufficiently high to permit some degree of predation efficiency. The salmon are probably invulnerable to visual-feeding piscivores at night, and also during daylight hours, when they remain in deep water. Narver (1970) rejects predation as a major determinant of the salmon migrations, on the grounds that very few trout were captured in purse seine sets: 30 sets made at night yielded only 10 rainbow trout and 11 lake whitefish *Coregonus clupeaformis*, compared with over 17,000 juvenile sockeye (Narver 1970,p. 286). However, there is no reason to expect the predators to be at the same depth as the sockeye at night, so that the actual number of predators present in the lake could be considerably higher than the observations suggest. As noted above, it does not take many large piscivores to impose a significant risk of mortality over an entire summer.

In Warner Lake, golden shiners migrate from the protected littoral zone into limnetic waters to feed at twilight, presumably thereby taking advantage of the antipredation window. The shiners' major limnetic prey, *Daphnia galeata*, undergo normal vertical migration, spending daylight hours in deeper water (> 4 m depth; Warner Lake is highly turbid, with Secchi disk readings of 4-5 m) and rising to within 1 m of the surface about 45 min after sunset (Table 5.3). The larger *Daphnia* ascend to the surface later than do the smaller *Daphnia*.

Why do *Daphnia* not spend the daylight hours at the surface, if shiners are absent? Hall et al. (1979, p. 1030) list 12 other species of fish that inhabit Warner Lake, many of which are also planktivorous, and it appears likely that *Daphnia* avoid exposure to these nonmigrating predators (see also Mittelbach 1981). Why then

Table 5.3
Vertical distribution of *D. galeata* in Warner Lake (Hall et al. 1979, p. 1035).

| time | 18:00 | 21:45 | 22:15 | 23:00 |
|---|---|---|---|---|
| light intensity (lx) | $> 1000$ | 20 | $< 0.1$ | $< 0.1$ |
| density of *Daphnia (/L)* | 15 | 17 | 36 | 38 |
| mean length of surface *Daphnia* (mm) | 0.93 | 1.05 | 1.25 | 1.27 |

do *Daphnia* not remain at depth until all risk of predation from golden shiners has passed? In fact, the largest *Daphnia* do adopt a more conservative migration strategy than do their smaller relatives (Table 5.3). Hall et al. show that the shiners elect larger prey when available.

Hall et al. (1979) conclude their article with the comments, "More detailed studies on the mechanism responsible for prey selection in relation to light, density, and prey size are needed to appreciate fully the importance of the precipitous changes in conditions that planktivorous fishes experience during the crepuscular feeding period. Such studies will reveal not only the nature of the coadaptation of plankton, planktivores, and piscivores, but also generate explicit hypotheses concerning foraging behavior." We would only add that mathematical models of diel migratory behavior should be included as an integral component of these studies. Such models can help to organize and explain existing observations, while leading to new hypotheses which themselves suggest further tests and studies. State variable models seem especially well suited to this task, because they can be closely tied to the underlying biology.

# 6 Parental Allocation and Clutch Size in Birds

One of the original uses of optimality thinking in behavioral ecology was David Lack's development of ideas about clutch size (Lack 1954, 1966, 1968). Lack argued that adult birds should not produce the maximum number of nestlings possible each year, but should produce the maximum number of young that can be successfully reared. That is, the *expected number of offspring* should be maximized. This prediction can be further modified, by including the effect of the current year's clutch on overwinter survival and future reproduction. Lack's ideas generated considerable empirical and theoretical work, and resulted in a large literature on the subjects of clutch size and parental allocation. (Examples are the papers of Klomp 1970, Johnsgard 1973, Burtt 1977, O'Connor 1978, Robertson and Biermann 1979, Drent and Daan 1980, Kacelnik 1984, Nur 1984a,b, Slagsvold 1984, Winkler 1985 Ekman and Askenmo 1986, Nur 1986, Slagsvold 1986, Boyce and Perrins 1987). In this chapter, our objective is to illustrate how the framework of dynamic modeling can be used to model parental allocation and clutch size, and to generate hypotheses for field testing. The general framework that we develop could be adapted to a variety of field or experimental situations.

We begin the chapter with a brief review of the avian life cycle. We then develop a simple model of parental allocation of food, and use this model to predict the optimal clutch size in a single year. This model, which involves state variables that characterize parent and nestlings, can be used to predict the optimal clutch sizes and optimal parental allocation. We next show how to couple the simple model with a model of parental overwinter survival, to obtain a multi-year model of parental allocation and clutch size. We discuss the difficulties (both conceptual and computational) that arise in the multi-year model. In the last section, we consider tests of these models using hypotheses generated by the solution of the dynamic programming equations.

In general, the life cycle pattern of many species of northern songbirds can be described as follows:

*February/March*: Males establish territories; females are attracted to the territories; nests are built.
*March–May*: Eggs are laid; the young hatch and are reared.
*May/June*: A second clutch may be laid and reared.
*July/August*: Molting occurs.
*August–October*: Food reserves are collected for overwinter survival.

For most birds, one egg is laid per day, so that large clutches imply a decrease in time available later in the season for other activities. Typical timing for the interval between egg laying and independence of offspring is about 12 days from eggs to hatching, about 12 days from hatching to fledging, and between 4 and 7 days between fledging and independence, after which the parents ignore the young. The nature and likelihood of a second clutch varies between species. Some species almost always lay a second clutch, some never do so, while in others a second clutch is occasionally laid. Dynamic modeling can help provide insight about the relationship between environmental and physiological variables and the likelihood of a second clutch.

## 6.1 A Single-Year Model of Parental Allocation and Clutch Size

To begin, we describe a model for the behavior of a parent bird and its allocation of food resources to its nestlings (our model is similar to the one in Mangel and Clark 1986). In this model only a single year is considered. We assume that the brood consists of $N$ nestlings, that all nestlings are treated identically, and that it is sufficient to consider the state variable of only one parent. Consequently, we have only two state variables in the model:

$X_p(t)$ = parental state variable (body weight or energy reserves at the start of period $t$)

and

$X_n(t)$ = state variable.for any one of the $N$ nestlings at the start of period $t$. (6.1)

In addition, we assume the following conditions for this simple model:

(1) The nestlings reach independence at period $T$. Thus, the interval of interest is $1 \le t \le T$. Here $t = 1$ (the first period) corresponds to the calendar time at which all nestlings have hatched and $t = T$ corresponds to the period in which the nestlings reach independence. (We ignore the possible dependence of $T$ on $N$ at this time.) If there are $T_d$ periods per day and the interval between hatch and independence is $D$ days, then $T = DT_d$.

(2) The parent decides at the start of each period whether or not to forage. When the parent is gone from the nest, there is a probability $\beta_n$ that the nestlings are killed by a predator and a probability $\beta_p$ that the parent is killed by a predator while foraging.

(3) The probability that a food item of type $i$ with energy value $Y_i$ is found when foraging is $\lambda_i$, $i = 1, 2, \ldots, I$. The probability of finding no food in a single period is $\lambda_0 = 1 - \sum_{i=1}^{I} \lambda_i$, with the corresponding value $Y_0 = 0$. If the parent bird finds food when foraging, then there is an additional decision concerning allocation. The parent can allocate a fraction $\phi$ of the food it discovers to the nestlings. In that case, the parent eats $1 - \phi$ of the food discovered, and each nestling receives $\phi/N$ of the food discovered.

(4) Daily metabolic costs are averaged over daytime and nighttime periods (this restriction is removed later in this section). Thus, let $\alpha_p$ denote the per-period metabolic cost of the parent when it stays at the nest and let $\alpha_f$ denote the per period cost of foraging. The per-period metabolic cost of the nestling, denoted by $\alpha_n(t)$, is a function of time and possibly of the number of nestlings (since the nestlings may warm each other, fight with each other, and otherwise interact). Similarly, the parental capacity is denoted by $C_p$, and the capacity of a nestling, denoted by $C_n(t)$, is also a function of time. A specification for the time dependences might be (see Figure 6.1) that $\alpha_n(1) = 0.05\alpha_p$ and $C_n(1) = 0.05C_p$ and both of these rise linearly to the parental value by period $T/3$. (For real birds, such as the starling *Sturnus vulgaris*, the parental metabolic rate and capacity are reached at about 7 days of age—A. Kacelnik, personal communication.) The critical (starvation) values for parental and nestling state variables are both 0.

For a given clutch size $N$ we define the function

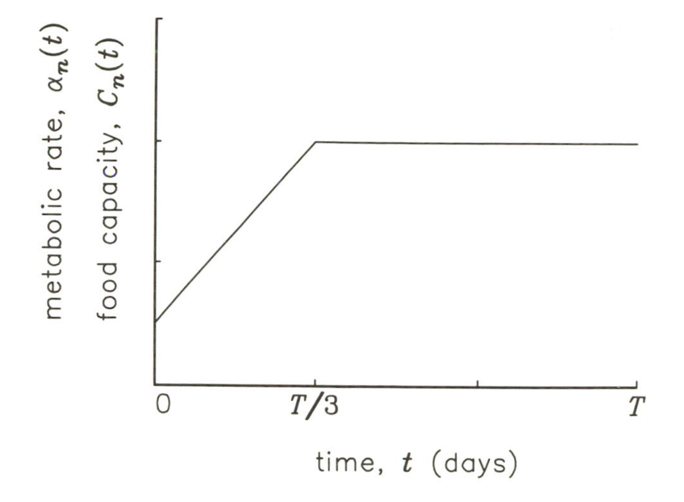


Figure 6.1 A hypothetical growth curve for nestling metabolic rate and capacity. The important feature of the growth is the rise to the adult level from a relatively low initial value.

$$F(x_p, x_n, t, T, N) = \max \Pr \{\text{the nestlings survive from period } t \text{ to independence a period } T \text{ given that } X_p(t) = x_p \text{ and } X_n(t) = x_n, \text{ brood size is } N\}. \tag{6.2}$$

In this equation, the maximization is taken over the parent's decisions about (i) whether or not to forage, and (ii) the fraction of any forage found which is allocated to the nestlings. With this definition, the quantity

$$C(x_p, x_n, N) = NF(x_p, x_n, 1, T, N) \tag{6.3}$$

is the maximum expected number of offspring reared by the parent, given that the initial value of the parental state variable is $x_p$ and the initial value of the nestling state variable is $x_n$. The choice of $N$ that maximizes $C(x_p, x_n, N)$ for particular environmental parameters can thus be interpreted as the "optimal clutch size," for this particular model.

As before, we begin the analysis by deriving the end condition for the function $F(x_p, x_n, t, T, N)$. When $t = T$, if the nestling state variable exceeds the critical value (which is set equal to 0), then the nestlings have survived. Hence, for all values of $x_p$ we have the condition

$$F(x_p, x_n, T, T, N) = \begin{cases} 1 & \text{if } x_n > 0 \\ 0 & \text{if } x_n = 0. \end{cases} \tag{6.4}$$

We must also consider what happens if the parent dies ($x_p = 0$) before $t = T$. Then the parent can no longer provide either nourishment or protection to the nestlings, but it may nevertheless be possible for the nestlings to survive to independence. Survival requires that two events occur: (i) there must be no predation at the nest between $t$ and $T$, and (ii) the nestling state variable must be sufficiently large. Combining these, we obtain

$$F(0, x_n, t, T, N) = \begin{cases} (1 - \beta_n)^{T-t} & \text{if } x_n > \sum_{s=t}^{T-1} \alpha_n(s) \\ 0 & \text{otherwise.} \end{cases} \tag{6.5}$$

In this equation, the summation of the $\alpha_n(s)$ represents the total metabolic needs of the nestlings between period $t$ and independence. We note that the assumption that the nestlings can survive without any parental care may be unreasonable for many birds since small nestlings are not likely to survive for any long period in the absence of a parent. In this case, Eq. (6.5) is replaced by the condition

$$F(0, x_n, t, T, N) = 0 \quad \text{for all } x_n \text{ and } t < T.$$

The dynamic programming equation for $F(x_p, x_n, t, T, N)$ can now be derived. The parental decision is either to remain at the nest (in which case the state variables corresponding to both parent and nestling decrease) or to forage (in which case the optimal allocation of forage must also be decided, if anything is found). The dynamic programming equation is

$$\begin{aligned} F(x_p, x_n, t, T, N) = \max\{&F(x_p - \alpha_p, x_n - \alpha_n(t), t+1, T, N); \\ &\beta_p(1 - \beta_n)F(0, x_n - \alpha_n(t), t+1, T, N) \\ &+ (1 - \beta_p)(1 - \beta_n) \sum_i \lambda_i \\ &\max_{0 \le \phi \le 1} F(x'_{pi}(\phi), x'_{ni}(\phi, t), t+1, T, N)\}. \end{aligned} \tag{6.6}$$

Table 6.1

Maximum probability of survival and optimal clutch size (from "Towards a Unified Foraging Theory" by M. Mangel and C. W. Clark, *Ecology* 1988, 67:1127–1138. Copyright ©1988 by The Ecological Society of America. Reprinted by permission.)

| Clutch Size $N$ | Prob. of Survival $F(5,1,1,T,N)$ | Expected No. of Survivors $C(5,1,N)$ | Prob. of Survival $F(5,2,1,T,N)$ | Expected No. of Survivors $C(5,2,N)$ |
|---|---|---|---|---|
| 1 | 0.63 | 0.63 | 0.68 | 0.68 |
| 3 | 0.55 | 1.65 | 0.60 | 1.80 |
| 5 | 0.34 | 1.70 | 0.42 | 2.10 |
| 7 | 0.05 | 0.35 | 0.10 | 0.70 |
| 9 | 0.001 | 0.009 | 0.006 | 0.054 |

Parameters are: $C_p = 10$, $\alpha_p = 2$, $\alpha_f = 3$, $\beta_n = 0.05$, $\beta_p = 0.025$, $Y_0 = 0$, $\lambda_0 = 0.15$, $Y_1 = 12$, $\lambda_1 = 0.85$, $\alpha_n(t) = 0.1 + \alpha_p(1 - e^{-0.3t})$, $C_n(t) = 0.1 + C_p(1 - e^{-0.3t})$, $T = 10$.

In this equation,

$$x'_{pi}(\phi) = \text{chop}(x_p - \alpha_f + (1-\phi)Y_i; 0, C_p)$$

and

$$x'_{ni}(\phi, t) = \text{chop}(x_n - \alpha_n(t) + \phi Y_i/N; 0, C_n(t)). \tag{6.7}$$

Here chop$(x; 0, C)$ is the chop function introduced in Chapter 2 (see Eq. 2.5). The first term on the right-hand side of Eq. (6.6) applies if the parent stays at the nest. The second and third terms apply if the parent forages. The second term corresponds to the case in which the parent is killed while foraging; the third term corresponds to the case in which both parent and nestlings survive foraging, and the parent must make the second decision of how much food to share with the nestling. The results of the computer solution of Eq. (6.6) are given in Table 6.1, which shows the maximum probability of survival of the nestlings, and the corresponding expected number of surviving fledglings. (Two different values of $x_n$ are used in Table 6.1, as a check that the computer program performs reasonably. As expected, the probability of survival increases with $x_n$ and decreases with $N$.)

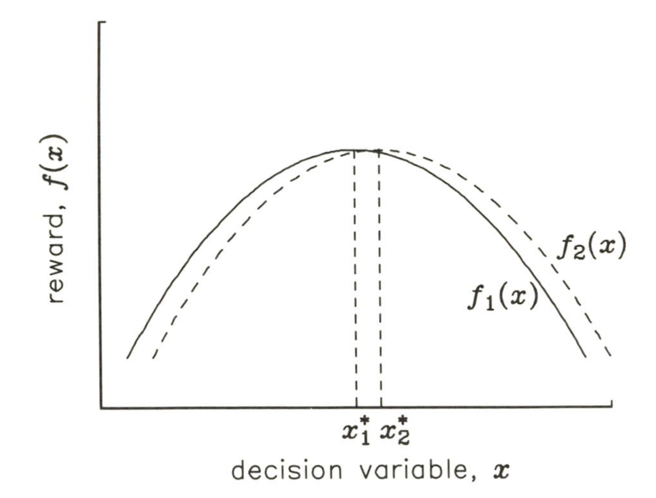


Figure 6.2 Nonrobustness of optimal decisions: The optimizing values of $x$ for $f_1(x)$ and $f_2(x)$ differ by 10%, but the penalty for a decision that is incorrect by 10% is approximately 1%.

From Table 6.1, we see that the optimal clutch in this model is $N = 5$. Note also, however, that clutches in the range of 3–5 give nearly the same value for the expected number of offspring. If this were a model of a real-world situation, with parameter values estimated from real data, the common-sense prediction would be that clutch size should range from about three to five eggs per clutch.

It is a commonly overlooked feature of optimization models that the computed optimand* is not very robust. Slight perturbations to the model or to its parameter values may result in relatively large changes in the value of the optimand (see Figure 6.2). Given that the model is inevitably a simplification and an approximation of the real world, it follows that quantitative predictions about optimal decisions may not be accurately matched by observations. Many arguments about the "correctness" of a given model have foundered on this simple point.

A behavioral model should not necessarily be rejected on the basis that observed behavior departs from the quantitative predictions. What is important is to check whether the value of the objective function (i.e. fitness) is seriously reduced by the appar-

* The optimand is the value of the independent variable ($x$, say) that maximizes the objective function ($f(x)$, for example).

Table 6.2
Optimal parental allocation

| Periods Until Independence $(T - t)$ | Optimal Allocation of Food When $x_p(t) = 5$ $x_n(t) = 2$ $N = 5$ |
|---|---|
| 1 | 1.00 |
| 3 | 0.85 |
| 5 | 0.85 |
| 7 | 0.80 |
| 9 | 0.55 |

ently suboptimal behavior. If so, then there are strong grounds for rejecting the model. Of course if observed behavioral patterns differ in major ways from model predictions, then the model is not of much use.

In Table 6.1, note that clutches of size one, or seven, would be severely suboptimal in terms of the expected number of surviving nestlings. If such clutches were frequently observed, the model would have to be placed under suspicion.

When solving the dynamic programming equation (6.6), we also obtain the optimal allocation of food as a function of parent and nestling state variables, clutch size, and time within the season. An example is shown in Table 6.2. The qualitative prediction from these results is that parents should allocate a higher proportion of food discovered to nestlings as the day of fledging approaches. (Which assumption of the model do you think is most involved in producing this prediction? What would you expect to happen if this were to be changed to a more realistic assumption?)

This relatively simple model leads to predictions about clutch size and parental allocation of food resources within the season. A straightforward elaboration of the model considers days and nights separately. Suppose that the $D$ days between hatching and independence consist of $T_d$ daytime periods in which the parent bird may forage, and one long nighttime period with no foraging. For example, a 24-hour day may be broken into 15 hours of foraging with foraging periods lasting five minutes, plus nine hours of night (see Figure 6.3).

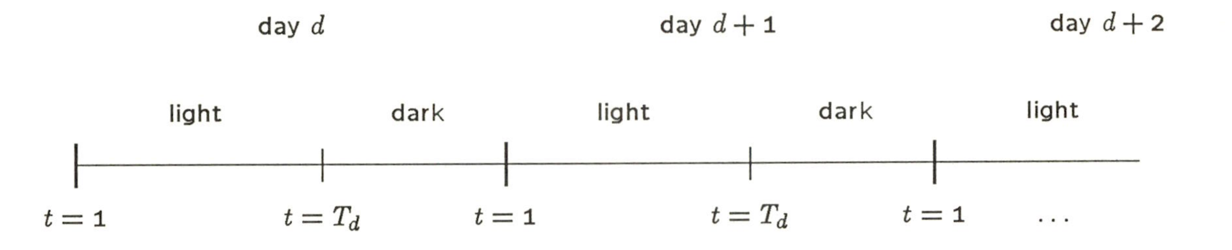


Figure 6.3 Division of days into daytime and nighttime periods, for use in the sequential coupling of the dynamic programming equations. If the state variables of parent and nestling are $x_p$ and $x_n$ respectively at period $T_d$ on day $d$, they are $x_p - \gamma_p$ and $x_m - \gamma_n(d)$ at period 1 on day $d+1$. This dynamic relationship allows one to couple the fitness function across days.

The corresponding function is now

$$F(x_p, x_n, t, d, T_d, D) = \max \Pr \text{ (nestlings survive to independence from period } t \text{ on day } d \text{ given that the state variables at the start of period } t \text{ on day } d \text{ are } X_p(t,d) = x_p \text{ and } X_n(t,d) = x_n). \quad (6.8)$$

The end condition is now that $F(x_p, x_n, T_d, D, T_d, D) = 1$ as long as $x_n$ exceeds the critical value (still assumed to be 0). The condition analogous to Eq. (6.5) is derived in a similar manner, or we can simply assume that the nestlings do not survive to independence if the parent is dead. For definiteness, let us choose the latter assumption.

The dynamic programming equation will be similar to Eq. (6.6), except that explicit treatment of the differences between daytime and nighttime metabolic costs and predation probabilities must be included. First consider what happens at period $T_d$ of day $d$. Suppose that the nighttime metabolic costs of parent and nestling are denoted by $\gamma_p$ and $\gamma_n(d)$ respectively and let $\bar{\beta}$ denote the probability of overnight predation of the nest. If predation does not occur, the state variables of the parent and nestling at the start of the next day are respectively $x_p - \gamma_p$ and $x_n - \gamma_n(d)$. (If either of these is not greater than the critical value, we chop at the critical value, still presumed to be zero.) Parent and nestlings survive the night if these exceed the critical levels. Thus, we obtain a "coupling condition" between different days:

$$F(x_p, x_n, T_d, d, T_d, D) = (1 - \bar{\beta}) F(x_p - \gamma_p, x_n - \gamma_n(d), 1, d+1, T_d, D). \quad (6.9)$$

This equation has the following interpretation: The maximum expected survival probability at the last period of day $d$, when the parental and nestling state variables are $x_p$ and $x_n$, is the same as the maximum expected survival probability at the start of day $d+1$, taking into account the possibility of overnight predation $(1 - \bar{\beta})$ and the reduction of the state variables due to metabolic needs overnight. The possibility of overnight death due to starvation (energy deprivation) is also included by the condition $F(x_p, x_n, t, d, T_d, D) = 0$ for $x_p = 0$ or $x_n = 0$.

The dynamic programming equation for parental behavior within days is still given by Eq. (6.6). The following pseudocode applies to the sequential coupling part of the computation.

(1) The current value of $F(x_p, x_n, t, d, T_d, D)$ is denoted by $F0(x_p, x_n)$ and the value "one period ahead" is denoted by $F1(x_p, x_n)$. Here "one period ahead" means period $t+1$ on day $d$ if $t < T_d$ and it means period 1 on day $d+1$ if $t = T_d$.
(2) Initialize by setting $d = D$ and $t = T_d$. Then $F1(x_p, x_n) = 1$ as long as $x_n > 0$ and $F1(x_p, 0) = 0$ for all values of $x_p$.
(3) Cycle through periods on day $d$ as $t$ goes from $T_d - 1$ to 1. Compute $F0(x_p, x_n)$ by using the dynamic programming equation (6.6). When this loop is finished $F0(x_p, x_n)$ contains the value of $F(x_p, x_n, 1, D, T_d, D)$.
(4) Replace $d$ by $d-1$. Set $F1(x_p, x_n) = (1 - \bar{\beta})F0(x'_p, x'_n(d))$, chopping at the critical value, if needed.
(5) If $d \geq 1$ return to step 3. Otherwise stop.

This pseudocode provides the numerical solution of the dynamic programming equation.

We will not present numerical results for this extended model; the model could be used, for example, to predict changes in behavior near dawn and dusk. McNamara et al. (1987) have used a similar model to explain song behavior of birds in spring.

## 6.2 A Multi-Year Model of Parental Allocation and Clutch Size

We now want to show how the model in the previous section can—in principle—be extended to include more than one breeding season. This will allow us to deal, in a dynamic framework, with the question of the tradeoff between reproduction in a single year and expected future reproduction.

A parent bird that raises a large brood in a particular breeding season may be less likely to survive over the winter than a bird that raised a smaller brood. For example, if there is a very strong dependence of overwinter survival on parental state at the end of the breeding season, then one might predict clutches far smaller than a single season optimum. The brood size in the current year must take into account overwinter survival by the parent and expected number and size of future broods. In a formal way, the idea is similar to but more complicated than the insect clutch size decisions studied in Chapter 4. In the case of insect oviposition decisions, the question is whether to use an egg at the current site, thus reducing future expected reproduction, or to save it. In each case, the tradeoff is between current and expected future reproduction. Kacelnik (1987) develops a static model for the multi-year clutch-size problem; here we develop a dynamic model.

To start, we consider some of the conceptual difficulties that can arise in a multi-year model. The first question concerns the appropriate measure of fitness for the case of overlapping generations. Total lifetime reproduction may not be an adequate representation of fitness in this context, since the timing of reproduction may also be important. In the stochastic case there may be no adequate single index of fitness (Tuljapurkar 1982). We will bypass these difficulties, and continue to use total lifetime reproduction as a measure of fitness.

Second, there is the problem of characterizing the manifold reproductive strategies that are available. In the previous section we considered strategies involving clutch size and food allocation. Some additional strategies that may be relevant in the multi-year situation include brood reduction, migration, food hoarding, and other possibilities. We will not treat such complications here, however.

Specifically, we shall assume that the parent bird's only decision is clutch size $N$ at the beginning of each year. Within-season allocation of food is determined by the solution of the single-year problem in the previous section.* Fitness is represented by expected total lifetime reproduction, over a time horizon of $Y$ years. Thus the lifetime fitness function is

$$F(x_p, y, Y) = \text{maximum expected lifetime reproduction for a parent bird starting the breeding season in year } y \text{ with state variable } x_p. \quad (6.14)$$

The expected reproduction resulting from a clutch of size $N$ is given by $C(x_p, x_n, N)$ as in Eq. (6.3), where $x_n$ (assumed constant) is the initial state of a typical nestling. Hence the dynamic programming equation for the multi-year model is:

$$F(x_p, y, Y) = \max_N \left[C(x_p, x_n, N) + E\{F(x_p', y+1, Y)\}\right]. \quad (6.15)$$

* Note that this is another instance of the method of sequential coupling.

In this equation, the maximum is taken over the size of the clutch in year $y$, and $x'_p$ is the value of the state variable at the start of year $y+1$, given that a clutch of size $N$ is bred in year $y$ and the optimal within-season allocation is used.

The second term on the right-hand side of Eq. (6.15) is the expected future reproduction. The expectation is taken over the value of the parental state variable $x'_p$ at the start of year $y+1$. To complete the specification of the multi-year model we need to specify the distribution of $x'_p$.

Suppose that the state of the parent bird at the beginning of the breeding season in year $y$ is $x_p$. If a clutch of size $N$ is bred, and the optimal allocation policy described in Section 6.1 is followed, then the state variable of the parent bird at the end of year $y$ is a random variable $\tilde{X}_p(N, y, x_p)$ that depends upon $N, x_p$ and environmental parameters. We can use forward iteration based on the single-year model to compute the probability distribution for $\tilde{X}_p$. Let $P(x)$ denote this distribution, so that

$$P(x) = \Pr\{\tilde{X}_p(N, y, x_p) = x\}. \tag{6.16}$$

Next, we must consider survival between seasons and—conditioned on survival—the value of the state variable of the parent bird at the start of season $y+1$, given that its state variable at the end of season $y$ was $x$. This probability, and the value of $x'_p$, will depend on $\tilde{X}_p$ and will be influenced by stochastic overwintering events. Let

$$\begin{aligned} f(z \mid x) = \Pr\{&x'_p = z \text{ in year } y+1, \text{ given that} \\ &\tilde{X}_p(N, y, x_p) = x \text{ in year } y\}. \end{aligned} \tag{6.17}$$

In particular, if $x_c$ denotes the critical value for the parent state variable, then

$$\begin{aligned} f(x_c \mid x) = \Pr(&\text{parent fails to survive the winter,} \\ &\text{given that } \tilde{X}_p(N, y, x_p) = x). \end{aligned}$$

We presume that the function $f(z \mid x)$ is estimated from field data (see the next section). The expectation in Eq. (6.15) can thus be expressed as

$$E\{F(x'_p, y+1, Y)\} = \sum_z F(z, y+1, Y) \Pr(x'_p = z)$$

$$= \sum_z F(z, y+1, Y) \times$$

$$\sum_x \Pr\left(x'_p = z \mid \tilde{X}_p(N, y, x_p) = x\right) \Pr\left(\tilde{X}_p(N, y, x_p) = x\right)$$

$$= \sum_z F(z, y+1, Y) \sum_x f(z \mid x) P(x). \tag{6.18}$$

Here the first equality follows from the definition of expectation, the second equality uses the Law of Total Probability, Eq. (1.11), and the third equality follows by substituting from Eqs. (6.16) and (6.17).

Equations (6.15) and (6.18) constitute a dynamic model for multi-year parental allocation and clutch-size decisions. Note that there is a dynamic model imbedded within a dynamic model, since in order to find $C(x_p, x_n, N)$ we must first solve the dynamic model in Section 6.1. After that, we solve Eq. (6.15). The algorithm for solving Eq. (6.15), once $C(x_p, x_n, N)$ is known, is similar to the algorithms employed in Chapters 2 and 4.

## 6.3 Hypothesis Generation and Testing Dynamic Behavioral Models

The models developed in Sections 6.1 and 6.2 can be field tested. Both models involve parameters (metabolic rates, clutch sizes, predation probabilities, and environmental parameters) that are readily measured, although predation probabilities may be more difficult to measure than the others. The model developed in Section 6.1 allows us to predict clutch sizes that would be observed in the field, for birds that have a short adult life expectancy. Although the model itself leads to the prediction of a specific optimal clutch size, we noted that a range of clutch sizes gives near-maximum fitness, so that the observed clutch sizes should be expected to fall within this range.

The model also leads to testable predictions about the allocation of food between the parent and nestlings. The data needed to test this model would be (i) metabolic rates of parent and nestlings, (ii) estimates of predation probabilities, and (iii) the probabilities of finding food and the distribution of energetic values of food items. In certain kinds of foraging experiments (e.g. Kacelnik 1984), the probability of finding food can be set equal to 1 and the distribution of energetic values of food items carefully controlled.

In that case, only metabolic rates and predation probabilities need to be estimated.

A slight modification of the model to include variable handling times for different energetic items could be used to generate testable predictions for experiments such as those of Kacelnik (1984) shown in Figure 6.4. In these experiments, Kacelnik studied the allocation of food by a parent starling when it had a brood of either three or seven nestlings. In these "central place" foraging experiments, the parent flew to a central place to obtain food. The lines marked N3 and N7 on Figure 6.4 show that the parent bird took the same amount of food from the central foraging site regardless of brood size. The allocation between nestlings (the lines B7 and B3) and the parent (the lines S7 and S3) strongly depends on brood size, however; a higher proportion of food was allocated to nestlings for large clutches than for small clutches. These results are qualitatively understandable on intuitive grounds, but quantitative interpretation would require a model of optimal food allocation similar to the models of this chapter.

To test the clutch size model developed in Section 6.2, we need to know the "input–output" relationships for survivorship over the winter and parental state at the start of the next breeding season. These input–output functions could presumably be measured by monitoring weights and suvivorships of birds over the winter. Once such data were available, we could use the model described in Section 6.2 to predict clutch sizes as a function of adult lifetime, across species of birds, for example.

The dynamic models developed in this chapter can thus be tailored to a specific field situation and used to generate predictions about clutch size and within-season parental allocation. Such predictions can be field tested. Although we must generally use numerical methods to generate hypotheses from the dynamic models, there is no reason that this should not be done if it leads to interesting and testable hypotheses.

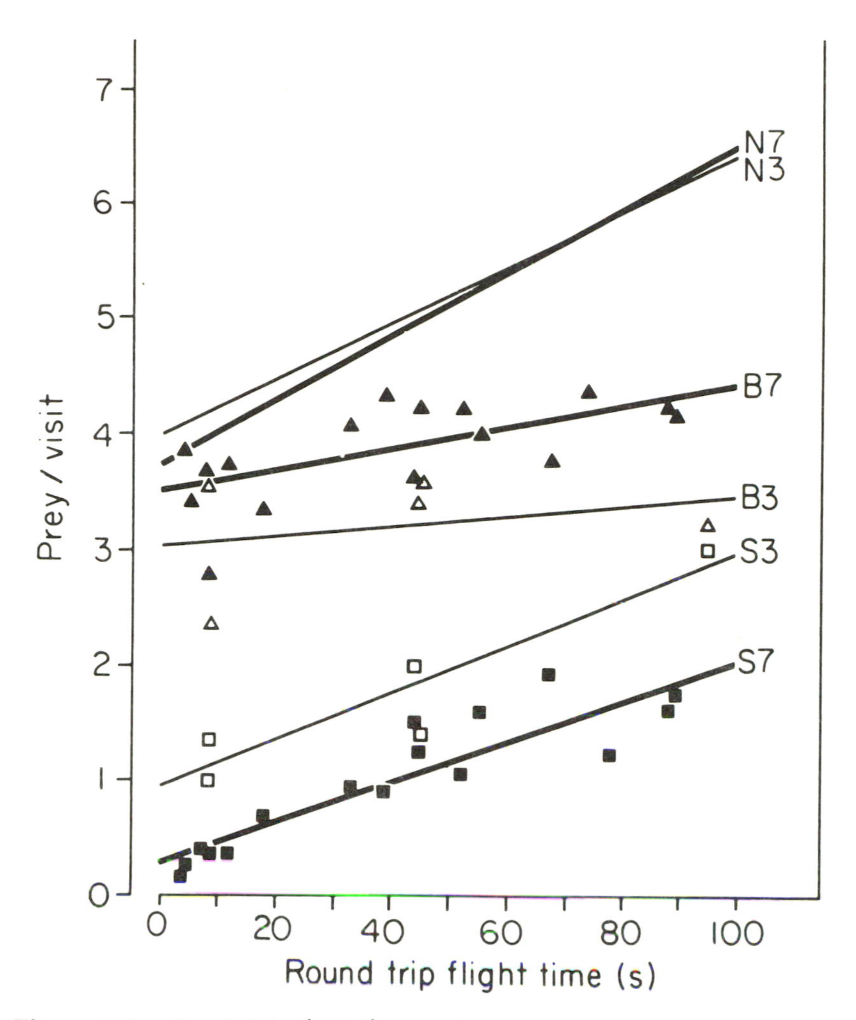


Figure 6.4 Kacelnik's (1984) experiments on parental allocation in starlings. The lines show the number of prey swallowed ($SC$) or carried to the nest ($BC$) and total number of items taken $NC = SC + BC$ as a function of travel time to the food patch, when the brood size was $C$, where $C = 3$ or 7. Note that the total number of items carried was essentially the same for each brood size, but the parental allocation differed. The dynamic models presented in this chapter can be tested using data such as these. (Reproduced with permission from the Journal of Animal Ecology 1984.)

# 7 Movement of Spiders and Raptors

The models developed in this chapter are motivated by two case studies. First we consider the choice of foraging mode of orb-weaving spiders in different habitats, drawing heavily on the work of Caraco and Gillespie (1986), Gillespie and Caraco (1987), and Wise (1975, 1979, 1983). The process of formulating the model leads to an understanding of what at first appear to be puzzling and enigmatic observations. The second case study concerns the natal dispersal and population dynamics of birds of prey. The model that we develop is motivated by work on the movement of the sparrowhawk *Accipter nisus* (see Marquiss and Newton 1981, Newton 1986, Newton and Marquiss 1983), and population regulation of the tawny owl *Strix aluco* (Southern 1970). In each case, we do not attempt to fit the model to a particular biological situation. Rather, in the spirit of Chapter 6, we develop a general model that could be tailored to a variety of different biological situations and use it to obtain general insights.

The issues considered in this chapter have wide generality. For example, a considerable literature exists on competition for breeding sites and movement in salmonids.* The study by Jenkins (1969) is especially noteworthy for its completeness and breadth in description of social structure, habitat selection, and movement in trout. Brown (1969) reviews territoriality and population in birds in general, and Greenwood, Harvey and Perrins (1979) discuss the causes and consequences of natal dispersal in the great tit *(Parus major)*. Huey and Pianka (1981) discuss the ecological consequences of foraging mode.

Natal dispersal has been studied in a number of mammalian species, e.g. lions (Hanby and Bygott 1987), marmots (Armitage 1986), and squirrels (Holekamp 1986). The movement and dispersal of the northern elephant seal (*Mirounga angustirostris*) during the last 20 years has been described by Le Boeuf and his colleagues

* E.g. Allee 1981, Bachman 1981, Cunjak and Power 1986, Hayes 1987, Jenkins 1969, Johnston 1981, Miller and Brannon 1981, Solomon 1981.

(see Le Boeuf and Mate 1978, Le Boeuf and Panken 1977). The elephant seal was nearly driven to extinction by hunters during the nineteenth century. Subsequent protection of this species led to the population increasing, and the areas in which the elephant seal breeds have also increased. At first additional islands off the coast of Mexico and California were colonized; ultimately the seals began also to breed on the mainland of California, at Año Nuevo Point. Models such as those developed in this chapter could be used to describe the movement of seals to new breeding grounds as the population increases.

## 7.1 Movement of Orb-Weaving Spiders

In this section we develop a dynamic, state variable model for the orb-weaving spider *Tetragnatha elongata*. This spider was studied in two different habitats by Gillespie and Caraco (1987), who also developed behavioral models. We first briefly describe the natural history and biology of the spiders, then introduce the dynamic model, and finally show how the process of dynamic modeling leads to improved understanding of the experimental results. In particular, we do not report any computer calculations in this section, since by the time the dynamic model is developed, the experimental puzzles are nearly completely understood. Numerical computations would simply be "icing on the cake"!

The first habitat used by the spiders was a wooded section of a creek (which we call the creek, denoted by $C$), and the second habitat was the border of a lake (which we call the lake, denoted by $L$). Prey items in the two habitats were essentially the same, consisting mainly of dipterans. Gillespie and Caraco (1987) obtained detailed measurements of prey availability and capture rates. They found that 61% of the insects hitting a web were actually captured and that the spiders consumed about 82% of the mass of a captured insect. Using sticky traps, they estimated prey capture rates as follows.

| | Prey Capture Rate (mg/day) | | |
|---|---|---|---|
| Site | Mean | Standard Deviation | Coeff. of Variation |
| $C$ | 1.74 | 0.3 | 0.17 |
| $L$ | 12.5 | 1.72 | 0.11 |

The results shown in this table indicate that the prey arrival rate at the creek was both much lower and slightly more variable than the prey arrival rate at the lake. Gillespie and Caraco estimated that in order to maintain its weight, a spider must consume about 0.9 mg of prey per day. They also estimated that an egg sac corresponds to consuming 45 mg of prey, and that the breeding season lasts 50 days. Thus, in order to produce an egg sac, a spider must consume at least $0.9 \times 50 + 45 = 90$ mg of prey over the 50 days.

Gillespie and Caraco measured the movement of spiders between web sites in each habitat. Spiders at the creek changed sites at 0.056 changes per day, whereas spiders at the lake changed sites at 0.753 changes per day. If we view these as the parameters of a Poisson process (see Appendix 1.1), then the mean residence time at the creek is $1/0.056 = 17.9$ days per site and at the lake the mean residence time is $1/0.753 = 1.3$ days per site. Gillespie and Caraco thus refer to the creek spiders as "sit-and-wait" predators and to the lake spiders as "mobile" predators. They also found a "transference of phenotypic plasticity" in which spiders originally at the creek, when moved to the lake, adopted the behavior of lake spiders (0.81 changes per day) and vice versa (lake spiders moved to the creek made 0.106 changes per day). The sites were sufficiently far apart that spiders did not move from creek to lake or vice versa by themselves. Gillespie and Caraco also measured the distances moved when a spider changed sites. At the creek, in changing sites a female moved about 4.4 m; at the lake she moved only about 1.6 m. However, during the process of building a new web, a spider moves a total distance of about 22.4 m, so that the effort associated with changing sites is relatively small in comparison to the effort in building a new web.

We now come to the main behavioral puzzle. Why do spiders at the creek move less frequently than spiders at the lake? The observation appears to be counterintuitive. Why don't spiders at the lake stop moving once they find a site with a high prey arrival rate, and shouldn't spiders at the creek keep moving, looking for sites with high prey arrival rates?

Caraco and Gillespie (1986) developed models for the computation of the probability of reproductive failure of sit-and-wait vs. mobile predators. They modeled prey capture at a particular web site as a Poisson process with parameter $\lambda$, assuming that $\lambda$ itself has a gamma distribution, and were able to determine when the strategy of sit-and-wait is preferable to the strategy of ranging widely, using the criterion of reproductive success. Gillespie and Caraco (1987) also developed a stochastic dynamic programming model in which fitness is defined in terms of prey interarrival times at sites. Thus maximizing fitness corresponds to choosing the site with the smallest prey interarrival time. This model should, at least in spirit, correspond to a stochastic version of a principle of rate maximization (here the rate being the rate of energy intake). The solution of their stochastic dynamic programming equation, however, showed that the mobility of the spiders should be independent of habitat quality. This is contrary to the observations. Finally Gillespie and Caraco developed a static model in which the objective is to minimize the probability of reproductive failure. The predictions of this model agreed with the observations.

We now show how to develop a stochastic dynamic state variable model of these field observations. Instead of choosing the probability of successful reproduction as the objective function, we choose the expected number of egg sacs (i.e. expected number of offspring) produced over the breeding season of $T = 50$ days. Gillespie and Caraco state that female *T. Elongata* can produce more than one egg sac during the breeding season. Wise (1975, 1979, 1983) gives further evidence for the prediction of more than one egg sac in response to increased food abundance.

The model that we now develop could be extended to include learning by the spiders about quality of sites, but we will not pursue that extension here. (In Section 9.1 we show how to treat such a model with learning.)

We choose the state variable to be $W(t)$, the weight of the spider at the start of day $t$, and assume that the corresponding critical value is $w_c = 0$. For the dynamics of the state variable, we assume:

(1) At the start of each day, the spider is at a web site. The $i$th type of site is characterized by a random variable $Y_i$ representing daily weight gain. Let $\Pr(Y_i = y_j) = p_{ij}$.

(2) The spider can choose either to remain at the current site or move to a new site. With probability $\lambda_i$ it arrives at a new site where the prey arrival distribution is the $i$th type. The chance of predation is $\mu$ if the spider chooses to move, but there is no chance of predation if the spider stays at the current site.

(3) If the spider stays at the current site, it receives a random increment in weight given by $Y_i - \alpha$, where $\alpha$ is the

daily metabolic cost. If the spider moves, it obtains no food during the day on which it moves, but incurs a metabolic cost $\alpha_m$.

Define the lifetime fitness function as

$$F(w,i,t,T) = \text{maximum expected number of egg sacs produced during the period between } t \text{ and } T\text{, given that } W(t) = w \text{ and that the prey arrival rate at the current site is type } i. \tag{7.1}$$

Using $k_e$ as the conversion factor of body weight to egg sacs gives the end condition

$$F(w,i,T,T) = k_e w. \tag{7.2}$$

This choice of end condition uses the assumption that the spider can convert a fixed fraction of its body mass into egg sacs. If there is a minimal nonconvertible residue $w_r$, then the right side of Eq. (7.2) would be replaced by $k_e(w - w_r)$. All egg sacs are assumed to be produced at the end of the season, but alternative assumptions could also be modeled.

The dynamic programming equation for $F(w,i,t,T)$ is readily derived since the only options that the spider has are to stay at the current site, in which case the dynamics of the state variable are

$$W(t+1) = W(t) - \alpha + y_j \text{ with probability } p_{ij} \tag{7.3}$$

or to move to a new site in which case the dynamics of the state variable are

$$W(t+1) = \begin{cases} W(t) - \alpha_m & \text{with probability } (1-\mu) \\ 0 & \text{with probability } \mu \end{cases} \tag{7.4}$$

and the new site is of type $k$ with probability $\lambda_k$. The values of $W(t+1)$ in Eqs. (7.3), (7.4) are "chopped" at the critical value 0 if needed. Note that no upper limit is placed on the spider's body weight in this model. The dynamic programming equation thus becomes

$$F(w,i,t,T) = \max\left( \sum_j p_{ij} F(w - \alpha + y_j, i, t+1, T); \right.$$
$$\left. (1-\mu) \sum_k \lambda_k F(w - \alpha_m, k, t+1, T) \right). \tag{7.5}$$

The first term on the right-hand side of Eq. (7.5) corresponds to the spider remaining at its current site and the second term corresponds to seeking a new site. Again, the chop at $w = 0$ is implicitly understood in Eq. (7.5).

In order to solve the dynamic programming equation (7.5), we must know the distributions $\{\lambda_i\}$ and $\{p_{ij}\}$. In principle, these distributions could be estimated from sampling using sticky traps, as Gillespie and Caraco did, but we will not discuss the necessary experimental design. (The data given by Gillespie and Caraco (1987) are not sufficient for the estimation. We note that in formulating a dynamic model for the spiders, we have identified experimental variables $(\lambda_i, p_{ij}, \mu)$ that need to be measured for a more complete understanding of the problem.)

Since the necessary parameter estimates are not available, we will not attempt to solve the dynamic programming equation (7.5). We believe, however, that formulating the problem in a dynamic context suggests a qualitative answer to the puzzle raised earlier about the movements of the spiders. When compared to creek sites, lake sites are both highly productive (with a mean prey capture rate about seven times higher) and more variable (with a variance in prey capture rate about 33 times higher). In Figure 7.1, we have schematically illustrated the two distributions for prey capture rates, treating them as continuous variables and highlighting the difference in variances. We also show the hypothetical value of prey capture rate, $y_1$, needed to produce a single egg sac.

Let us assume that at the start of the season spiders are randomly allocated to different site types. After just a few site changes, a spider at the creek will arrive (on the average) at a site in which the prey capture rate exceeds $y_1$, with little chance of finding a better site. On the other hand, a spider at the lake will essentially always be at a site with a prey capture rate sufficient to produce an egg sac, but also will generally have a non-negligible probability of finding a better site, especially if the tail of $f(y)$ is long. Thus, if the probability of predation while moving is small, we would predict that after the first few days of the season, spiders at the creek will move very little, but spiders at the lake will continue moving. If the probability of predation while moving is less than the probability of predation if not moving, then movement at the lake will be even more accentuated. We thus predict more movement at the lake than at the creek, and this provides

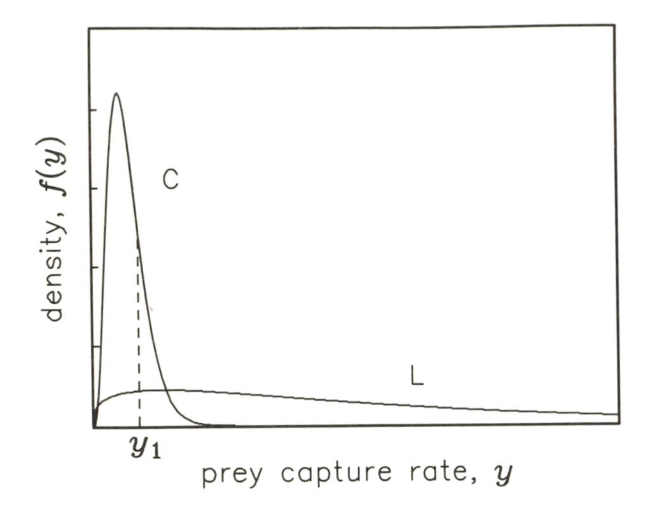


Figure 7.1 Schematic illustration of prey capture rates. Here $f(y)$ is the probability density that a creek $(C)$ site or lake $(L)$ site has prey capture rate $y$. The hypothetical prey capture rate required for the production of one egg sac, denoted by $y_1$, is also shown. The curves are drawn in an exaggerated fashion, to emphasize the more than 30-fold difference in variance between creek and lake sites.

at least a qualitative understanding of the results. We encourage the reader to code the dynamic programming equation and experiment with the effect of different probability distributions for $\{\lambda_j\}$ and $\{p_{ij}\}$ and different values of $\mu$ on the movement of the spiders in each habitat.

## 7.2 Population Consequences of Natal Dispersal

Figure 7.2 shows Newton's (1983) data on the dispersal of sparrowhawks from their breeding territory (also see Marquiss and Newton 1981, and Chapter 20 of Newton 1986). Newton reports that as the sparrowhawks move from their natal territory, their breeding success decreases. A natural tension thus exists between dispersal and reproductive success. In this section we develop a dynamic model for movement from a birth territory, and study the population consequences of individual natal dispersal. We show how a general model connecting natal dispersal, individual

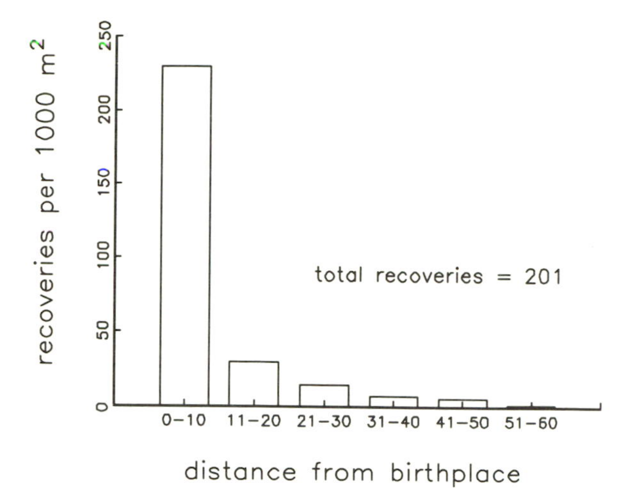


Figure 7.2 Natal dispersal of sparrowhawks (Newton and Marquiss 1983). The data show the locations of sparrowhawks recovered in the breeding season, in relation to the natal territory.

behavior, and population phenomena can be developed, without attempting to fit the model to any particular data. For example, although our model could be fitted to Figure 7.2 (or used to generate a comparable figure), we do not do so. Instead, we use the model to analyze qualitative features of population consequences, where a "population" consists of a collection of optimally behaving individuals.

De Jong (1979, 1982), Kiester and Slatkin (1974), and Patterson (1980) developed models for the effect of dispersal pattern on fecundity and the effect of spatial distribution on the dynamics of a population. These are "whole population" models, rather than individual behavioral models. Our approach is different: we model individual behavior and then study the population consequences of such individual behavior. The work of Lomnicki (1978) is closest to the approach in this chapter. Lomnicki assigns arbitrary ad hoc levels of food to each individual, rather than taking an approach in which the food obtained by individuals is computed by considering individual behavior. Even so, Lomnicki's paper is similar to our approach in that it captures the main features of spatial and temporal heterogeneity of resources and unequal partitioning of resources (see also Lomnicki 1988).

Our dynamic model uses the following assumptions:

(1) Breeding pairs maintain fidelity throughout their lives, so that it is sufficient to consider a single state variable characterizing energy reserves of the female. Once a pair moves from its natal territory, it does not return.
(2) Brood size and fledging date are independent of the distance from the natal territory. However, individuals breeding in the territories of their birth are assumed to have higher probabilities of successfully fledging young, and the probability of successful fledging decreases as the distance from the natal territory increases (Newton 1986).
(3) No breeding occurs after maximum age $Y$ (measured in years). Each year is broken into a breeding season lasting $T$ periods and a long "overwintering" period. Only one clutch per year is laid, and if a pair is successful it stays at the breeding site until period $T$, at which time the pair and its fledged young prepare for winter. Yearling overwinter survival is less than adult overwinter survival, and both clutch size and survival rates depend upon the state of the parent.
(4) Access to a breeding site in a territory is determined by a "lottery." If a territory contains $N_S$ breeding sites and $B$ breeding pairs, then

$$\begin{aligned}&\text{Pr (the pair gets a breeding site, given that } B \text{ other}\\&\qquad\qquad\text{breeding pairs are present)}\\&= p_S(B) = \min\left(1, N_S/(B+1)\right) \qquad (7.6)\end{aligned}$$

Territories have the same number of potential breeding sites, so that $N_S$ is constant across territories. (For the sparrowhawks, $N_S$ is typically 1, but it need not be for other species, since the definition of a territory is somewhat arbitrary.) A breeding pair that does not obtain a site can choose either to remain in the territory, not to breed in the current year and try to obtain a site in the next year, or to move to an adjacent territory where it may obtain a site but has a lower chance of nest success.

The dynamic model requires two time variables, age of the hen ($y$) and time within the season ($t$). The state variables are then

$X(t;y)$ = energy reserves of a hen of age $y$ at the beginning of period $t$ in the breeding season

$D(t,y)$ = distance the hen of age $y$ is from her natal territory at the beginning of period $t$

$B(t)$ = number of other breeding pairs in the territory at the beginning of period $t$. (7.7)

We define the lifetime fitness function by

$F(x,y,d,b,t,T)$ = maximum expected lifetime reproduction for an individual born in territory 1, currently in year $y$ of its life, with state variables $X(t,y)=x$, $D(t,y)=d$, and $B(t)=b$. (7.8)

The end condition associated with this fitness function is then

$$F(x,Y,d,b,T,T)=0. \tag{7.9}$$

In order to derive the dynamic programming equation that characterizes $F(x,y,d,b,t,T)$, we consider the possible behaviors and their consequences separately. We then combine the separate behaviors and their values into the dynamic programming equation. The different behaviors are (i) not to attempt to breed in the current territory and thus prepare for winter, (ii) to attempt to breed in the current territory, or (iii) to move to another territory if unsuccessful at breeding in the current territory.

We will specify the dynamics of the state variable $X(t,y)$ as each possible behavior is considered. To specify the dynamics of the state variable $D(t,y)$, we measure the distance in increments of 1 "territory unit." We then have, for within season,

$$D(t+1,y)=\begin{cases} D(t,y) & \text{if pair stays in its current territory} \\ D(t,y)+1 & \text{if pair moves to an adjacent territory} \end{cases} \tag{7.10}$$

and for between season,

$$D(1,y+1)=D(T,y). \tag{7.11}$$

That is, the pair's territory at the start of the next season is the same as its territory at the end of the current season.

*Preparing for winter.* We assume that preparing for the winter entails moving to an "overwintering site," and that the state variable is decreased by an amount $\alpha_m$ associated with the cost of moving. (Sparrowhawks actually do not physically move to an overwintering site, in which case we would simply set $\alpha_m = 0$.) The main point is that the pair gives up attempting to breed and begins to prepare for winter, for example by building up reserves of body fat. A longer preparation time leads to an increased probability of survival over the winter and to an increased state variable at the start of the next breeding season.

To characterize what happens over the winter, we specify an input–output relationship for the state variable (as in the previous chapter) and a probability of survival. Thus let

$$x_w(x,t) = \text{value of the state variable } X(1, y+1) \text{ given that } X(t,y) = x \text{ and the pair began preparing for winter at period } t. \tag{7.12}$$

In general, $x_w(x,t)$ will be a random variable.

Overwinter survival is characterized by a probability

$$p_{ws}(x,t) = \text{Pr (breeding pair survives over the winter, given that it begins preparation for winter in period } t \text{ and that } X(t,y) = x). \tag{7.13}$$

Let $V_m(x, y, t, d)$ denote the value (measured in expected future reproduction) of beginning to prepare for winter in period $t$, when $X(t,y) = x$ and $D(t,y) = d$. This value is

$$V_m(x, y, t, d) = p_{ws}(x,t)E\{F(x_w(x,t), y+1, d, \tilde{B}, 1, T)\}. \tag{7.14}$$

In this equation, $\tilde{B}$ denotes the (random) number of other breeding pairs in the territory at the start of the next year, and the expectation is taken over $\tilde{B}$ and the distribution on $x_w(x,t)$. Both of these probability distributions would need to be measured to model a particular situation.

*Breeding successfully.* Consider a pair that attempts to breed during period $t$ and is successful in obtaining a breeding site. The probability that it gains a breeding site is given by $p_s(b)$ in

Eq. (7.6). We need then to model the clutch size and the fledging success. Thus, let

$$C(x,t,d) = \text{clutch fledged by a successful breeding pair with state variable } X(t,y) = x \text{ and } D(t,y) = d \tag{7.15}$$

and

$$p_f(x,t,d) = \text{Pr (member of a brood, fledged in a clutch when the state variables are } X(t,y) = x,\ D(t,y) = d, \text{ survives through its first winter)} \tag{7.16}$$

Let $\alpha_b$ denote the reduction in the parental state variable associated with breeding successfully.

The value of successful breeding is the sum of the expected reproduction in the current year plus expected future reproduction:

$$V_b(x,y,t,d) = C(x,t,d)p_f(x,t,d) + p_{ws}(x-\alpha_b,T) \\ E\{F(x_w(x-\alpha_b,T),y+1,d,\tilde{B},1,T)\} \tag{7.17}$$

In this equation the first term on the right-hand side corresponds to the expected reproduction in the current year and the second term on the right-hand side corresponds to expected future reproduction.

*Attempting to breed and failing to obtain a breeding site.* Finally consider the situation in which a pair attempts to breed but does not obtain a breeding site during period $t$. Its choices are to prepare for winter at the start of period $t+1$ or to move to a new territory (using the dynamics in Eq. (7.10)) and to attempt to breed there. We let $\alpha_n$ denote the cost of moving to a new territory and $V_{ns}(x,y,t,d)$ denote the value of future reproduction to a breeding pair that does not obtain a breeding site in the current territory. Then

$$V_{ns}(x,y,t,d) = \max\,(E\{p_s(\tilde{B}(d+1))\times \\ F(x-\alpha_n,y,d+1,\tilde{B}(d+1),t+1,T)\}; \\ p_{ws}(x,t)E\{F(x_w(x,t),y+1,d,\tilde{B},1,T)\}) \tag{7.18}$$

where $p_s(B)$ is defined in Eq. (7.6). In this equation $\tilde{B}(d+1)$ is the number of other breeding pairs in the next territory; it is treated as a random variable in the same way that $\tilde{B}$ is treated as a random variable.

*The dynamic programming equation.* It is now possible to combine the values associated with the three different behaviors into the dynamic programming equation. The optimal decision is to choose the larger of $V_m(x, y, t, d)$ and the expected value of reproduction at the current territory. Thus

$$F(x, y, d, b, t, T) = \max[V_m(x, y, t, d), \\ p_s(b)V_b(x, y, t, d) + (1 - p_s(b))V_{ns}(x, y, t, d)]. \tag{7.19}$$

If the first term on the right-hand side of Eq. (7.19) is the larger, then the optimal decision when $X(t, y) = x$, $D(t, y) = d$, and $B(t) = b$ is to prepare for winter. If the combination of $V_b(x, y, t, d)$ and $V_{ns}(x, y, t, D)$ is larger, then the optimal behavior is to try to obtain a breeding site. If one is obtained, then the pair breeds in period $t$, and prepares for winter during period $T$. If a breeding site is not obtained during period $t$, then the pair either moves to a new territory or to the overwintering site during period $t+1$. This movement is determined by the maximum term on the right-hand side of Eq. (7.18).

*Population consequences of individual behavior.* The population consequences of individual behavior can be studied by considering what happens in a population that begins with just a few breeding pairs. To be specific, suppose that we start in year 1 with two breeding pairs in territory 1 and define the population in any subsequent year as the total number of breeding pairs summed over all territories. The original breeding pairs will reproduce and their young will ultimately exhibit natal dispersal as the population grows. One way to study the effects of natal dispersal is first to solve the dynamic programming equation (7.19) and then to construct a Monte Carlo simulation of organisms following the optimal behavioral decisions.

In order to solve the dynamic programming equation, the functional forms of survival probabilities, clutch sizes, and distributions must be specified. For the sake of illustration, the forms shown in Table 7.1 were chosen. In this table $\mu_1$ through $\mu_6$ and the $\{Y_i, \lambda_i\}$ are parameters that must be specified and $C^*(x)$ is the clutch size corresponding to state variable $x$. We also need to

Table 7.1
Functional forms for the solution of the dynamic programming equation

| Quantity | Interpretation | Functional Form |
|---|---|---|
| $C(x,t,d)$ | clutch size | $C^*(x)\exp(-\mu_1 d)\exp(-\mu_2/(T-t))$ |
| $p_f(x,d)$ | probability of fledging | $\exp(-\mu_3/x)\exp(-\mu_4 d)$ |
| $p_{ws}(x,t)$ | probability of overwinter survival | $\exp(-\mu_5/x_w(x,t))\exp(-\mu_6/(T-t+1))$ |
| $x_w(x,t)$ | state variable at the end of winter | $\text{chop}(x-\alpha_m-\alpha_w+(T-t)Y_i;0,X_m)$ with probability $\lambda_i$ |

specify the distribution on both $\tilde{B}(d+1)$ and $\tilde{B}$. For the computations reported below each of these is assumed to have a uniform distribution on the interval 0 to $B_m$, where $B_m$ is a parameter. For field testing these models the distributions could be determined through population sampling.

To study population consequences of individual behavior, the Monte Carlo simulation begins with just two breeding pairs in territory 1 in year 1 and follows the movement and breeding of the original pairs and their descendants, assuming that they act optimally according to the dynamic programming equation. Figure 7.3 shows the results of such computations. The following parameter values were used: $Y = 4$, $T = 4$, $B_m = 6$, $N_s = 2$, $\alpha_m = 1$, $\alpha_b = 2$, $\alpha_w = 2$, and a clutch structure in which $C^*(x) = 1$ for $x = 2,3,4,5,6$ and $C^*(x) = 2$ for $X \geq 7$, with a maximum value of $X = 9$. Three values of $Y_i$ and $\lambda_i$ were used: $Y_1 = 2$, $Y_2 = 4$, and $Y_3 = 6$, each with probability 1/3. The values of the $\mu_i$ were $\mu_1 = 0.1$, $\mu_4 = 0.8$, and $\mu_5 = 0.5$; all other $\mu_i = 0$. This figure exhibits growth of the population with fluctuations and an apparent "carrying capacity." The mechanism for population regulation in this example is the lottery competition for breeding sites, which is the operational origin of the carrying capacity.

These results show how a dynamic model in which individual behaviors are considered can be used to model population level

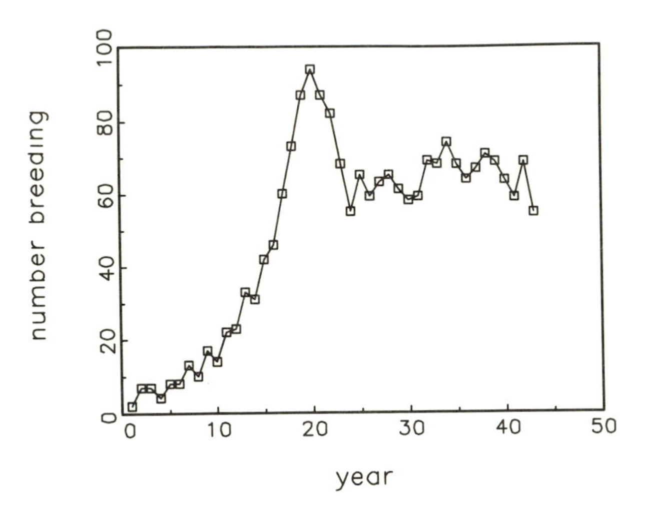


Figure 7.3 Population consequences of individual behavior. In year 1, two breeding pairs start in territory 1. This figure shows typical results of Monte Carlo simulation for the first set of $\{\mu_i\}$ described in the text. The total population (i.e. total number of breeding pairs summed over territories) is plotted as a function of time.

phenomena. A specific model of the sparrowhawk or tawny owl would involve modification of the model developed in this chapter, but is entirely feasible.

# III Additional Topics

*Our objective in this third part is to provide an overview and some general perspectives on dynamic modeling in behavioral ecology. In Chapter 8 we discuss the formulation and solution of state variable models in a more general way than in Chapter 2. Next we describe some additional useful techniques for the solution of the dynamic programming equations. Finally we discuss alternative modeling approaches, and examine their relationship to the state variable (or dynamic programming) approach.*

*In Chapter 9 we show how to extend the dynamic modeling methodology to include (a) learning about environmental parameters, and (b) interactions with other organisms. Including either of these aspects can lead to major complications in attempting to solve the dynamic programming equations.*

*Finally, we end the book with an Epilogue on perspectives in modeling, as applied to dynamic models in behavioral and evolutionary ecology.*

# 8 Formulation and Solution of State Variable Models

In this chapter we describe the general process involved in setting up a dynamic behavioral model. We presume that the reader has been through Chapters 1 and 2 and at least one of the applications chapters in Part II. This chapter concerns some of the intricacies of formulation and solution of dynamic state variable models. It casts the dynamic models into a general mathematical framework and shows how that framework is analyzed and used in computations. Alternative modeling approaches are discussed briefly in Section 8.4.

## Optimization Models

The essential components of an optimization model are (Oster and Wilson 1978):

(1) a state space;
(2) a set of constraints;
(3) a strategy set; and
(4) an optimization criterion.

In the case of a *dynamic* optimization model, we also need to specify

(5) state dynamics.

In a stochastic model, some or all of these components will involve probability distributions.

For example, the basic patch selection model discussed in Chapter 2 had the following components:

(1) State space:
$X(t)$ = energy reserves of the organism at the start of period $t$

(2) Constraints:
  $x_c \leq X(t) \leq C$ for all $t$
(3) Strategy set:
  choice of patch or activity $i$ ($i = 1, 2, \ldots, N$)
(4) Optimization criterion:
  probability of survival over $T$ periods
(5) State dynamics:
  if patch $i$ is chosen, then

$$X(t+1) = \begin{cases} \text{chop}(X(t) - \alpha_i + Y_i; 0, C) & \text{with prob. } (1-\beta_i)\lambda_i \\ \text{chop}(X(t) - \alpha_i; 0, C) & \text{with prob. } (1-\beta_i)(1-\lambda_i) \\ 0 & \text{with prob. } \beta_i \end{cases}$$

Once such a model has been constructed, the next step is to deduce the optimal strategy or strategies—in this case the optimal choice of patch—as a function of the current time $t$ and the current value of the state variable $X(t)$. Having thus constructed and "solved" the model, we are in a position to make either qualitative or quantitative predictions. Obtaining such predictions is, of course, the original purpose of modeling. The model's success is judged in terms of the degree to which its predictions concur with field or experimental data, or lead to new understanding of the phenomenon being studied. And if the model explains data that previous models failed to explain, then we can perhaps feel a sense of accomplishment in having advanced scientific understanding, *especially if the model itself is in some sense a convincing, even though simplified, description of the underlying biology.*

Although we discuss perspectives on modeling in more detail in the Epilogue, we wish to stress this last point, since it represents our philosophy as modelers. We believe that there are good ways and bad ways to build models. Simple models are, within limits, better than complex models; models which are biologically meaningful, and contain parameters that have biological significance, are better than abstract or mechanical models. Finally, of course, the acid test of any model is its ability to explain and predict natural phenomena, but we are not willing to sacrifice simplicity or realism in vain attempts merely to "improve the statistical fit" to data (Fagerström 1987).

In setting up an optimization model of some behavioral phenomenon, how does one go about specifying the necessary components? Clearly there is no unique answer to this question: modeling is a creative art rather than a rigidly defined routine. We learn by doing, and by observing the efforts of others. Remember, this is a book about modeling, rather than a book of models. In this chaper we hope to pass along a few hints for good modeling in behavioral ecology, based on our own experience.

The problems of specifying state variables, constraints, and dynamic laws are discussed in Section 8.1. Next the question of modeling fitness (which we always assume to be the basic optimization criterion, at least for individual decisions) is taken up in Section 8.2. In Section 8.3 we discuss some complexities of solving dynamic optimization problems. Section 8.4 is devoted to a brief survey of alternative modeling approaches. In the Appendix we show how alternative definitions of fitness in fluctuating environments can be treated in the dynamic, state variable framework.

## 8.1 Identifying State Variables, Constraints, and Dynamics

The applications in Chapters 3–7 suggest two simple rules for choosing state variables:

(1) Think deeply about the biology.
(2) Be parsimonious.

By a state variable we mean any characteristic of the organism that will affect its behavior. State variables typically vary over time, depending on what the organism does, and possibly also depending on random events. For example, if we are modeling foraging behavior, and growth is not considered to be very important, then a natural state variable would probably be either energy reserves or gut contents. Obviously the state of an animal cannot be fully represented in such a simple manner, but many behavioral phenomena may depend critically on energy reserves, so that neglecting other state variables may not be unrealistic as a first approximation. Anyway, including just one state variable may well be a noticable advance over previous models which were devoid of state variables entirely!

In many behavioral contexts, size may be an important component of lifetime fitness. In this case body weight or length becomes a primary choice for state variable. We might simplistically assume that food intake in excess of daily metabolic requirements immediately induces growth, but a more complex model would include gut contents, stored energy reserves, and body weight as the state variables. We would then also have to model the metabolic processes relating all these quantities. Quantities other than gut contents, energy reserves, or body weight may sometimes be more appropriate as primary state variables. For example, in Chapter 4 we discussed models for the oviposition strategy of parasitic insects; the basic state variable in this instance was the (potential) number of unlaid eggs. Other examples of state variables encountered in Chapters 3–7 include territory size and number of offspring being cared for. In Chapter 9 we will discuss learning models, in which the state of information possessed by the organism becomes an additional state variable.

We hope that the foregoing discussion indicates what we mean by "thinking biologically." In setting up a dynamic model, you are forced to ask yourself what are the most important state variables for the behavioral phenomenon under investigation. Often the answer is intuitively evident, and the modeling process can begin. Later you may decide to develop a more complex state model.

## Constraints

The specification of constraints on state variables is often straightforward. For example, if $X(t)$ represents gut contents or energy reserves, then the constraints

$$0 \leq X(t) \leq C \tag{8.1}$$

where $C$ denotes capacity, are almost automatic. Because of the prevalence of such interval constraints in our models, we introduced the useful "chop" function

$$\text{chop}(x; a, b) = \begin{cases} x & \text{if } a \leq x \leq b \\ a & \text{if } x < a \\ b & \text{if } x > b. \end{cases} \tag{8.2}$$

In a model with $W(t)$ = body weight as a second state variable, we would probably want to assume that gut or energy capacity is related to body weight. A simple specification would be

$$C = k_1 W(t)^{k_2}. \tag{8.3}$$

The parameters $k_1$ and $k_2$ could be estimated from allometric data (e.g. Schmidt–Nielsen 1984).

### Strategies

Next, a set of feasible strategies, or actions, must be specified. The set of actions to be included in the model is determined by the behavioral phenomenon under investigation. In a laboratory setting the action set can be controlled by experimental design, but the problem becomes much more difficult in field studies, because of the wide choice of actions usually available to an animal. Simplifying assumptions must therefore usually be adopted. This often works both ways—we typically restrict the set of actions to be considered (although the dynamic approach is much more flexible than static modeling in this respect), but at the same time allow actions that could not realistically be performed. For example, we often assume that the animal acts on perfect information regarding environmental parameters (but not regarding the future values of random variables), whereas in fact this is probably seldom the case. Questions of imperfect information are addressed in Chapter 9.

### State Dynamics

Dynamic transition properties are crucial for the construction of a state variable model. In this book we used discrete-time dynamic models exclusively. There are two main reasons behind this choice. First, discrete-time models are easy to formulate and to interpret biologically, relative to continuous-time models. This is especially true for stochastic models, which require the advanced mathematical techniques of stochastic differential equations in the continuous-time setting, but are more elementary in discrete time. Second, discrete-time optimization problems are easily converted into computer programs (via dynamic programming), from which numerical solutions can be obtained. You do not need to be an expert in the intricacies of optimal control theory in order to formulate and solve discrete-time stochastic dynamic optimization models on the computer.

The general modeling structure can be illustrated schematically as shown in Figure 8.1: the animal is in state $X(t)$ at the beginning of time period $t$; it undertakes an action $A(t)$, which affects its state $X(t+1)$ in the next period, and may also result in an immediate reward (e.g. reproductive output) $R(t)$. Both the reward and the transition to the new state $X(t+1)$ may be subject to random effects $w(t)$. The process is repeated at periods

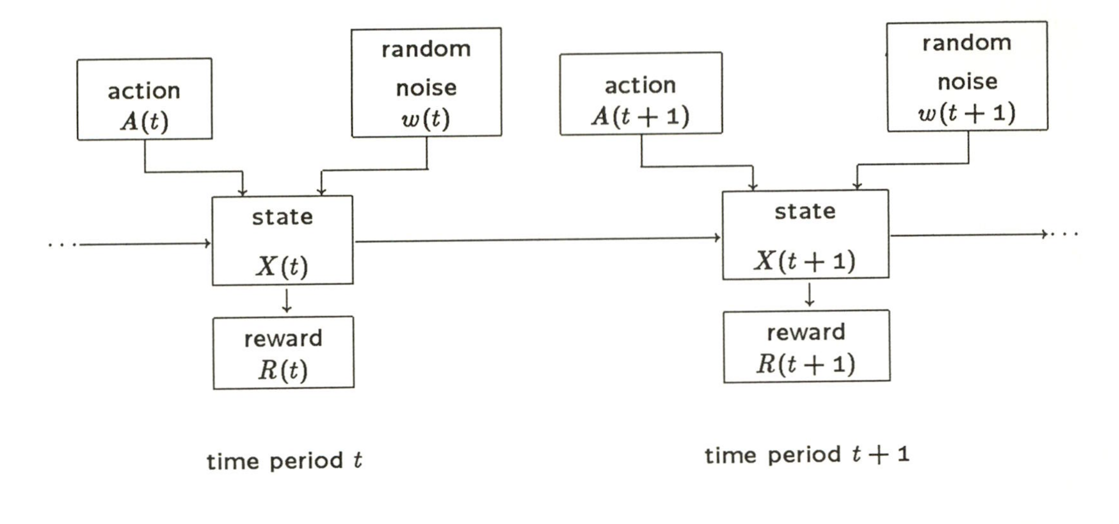


Figure 8.1 Schematic diagram of the evolution of a stochastic system subject to periodic choices of action $A(t)$ (from Clark 1985).

$t+1, t+2, \ldots, T$.

The state dynamics of this model can be expressed in the general form

$$X(t+1) = G(X(t), A(t), t, w(t)). \tag{8.4}$$

The function $G(x, a, t, w)$ includes state dynamics and constraints as appropriate. In Eq. (8.4), the state variable $X(t)$ may be multidimensional. The $t$th period reward can be expressed as

$$R(t) = R(X(t), A(t), t, w(t)). \tag{8.5}$$

But let us be more explicit. Assume both that the strategy set is finite (with strategies $A_i$, $i = 1, 2, \ldots, N$), and that the random variable $w$ is finite-valued, with

$$\Pr(w = w_{ij} \mid \text{strategy } i \text{ used}) = \lambda_{ij} \qquad (j = 1, 2, \ldots, N_i). \tag{8.6}$$

Then Eq. (8.4) can be written as

$$X(t+1) = G(X(t), A_i, t, w_{ij}) \quad \text{with probability } \lambda_{ij}. \tag{8.7}$$

In this notation, the dynamics of our model are specified by defining the function $G$ and the probability distributions $(\lambda_{ij})$, for each strategy $i$.

### Some Limitations

The stochastic optimization models that we are describing here are known technically as Markov Decision Processes. The theory of Markov Decision Processes has been extensively developed in the statistical and operations research literature (see Aoki 1967, Heyman and Sobel 1984). The characterizing Markovian feature of these models is the assumption that everything that transpires in the transition from period $t$ to period $t+1$ is determined by the state variable $X(t)$ only, and not on values of $X$ in previous periods $t-1, t-2, \ldots$. This is not a serious restriction since it can be shown that a model in which the state in period $t+1$ depends on $n$ previous periods ($n$ fixed) can be reformulated as a Markovian model. This is accomplished simply by defining a new model with a vector of state variables $Y(t)$ given by

$$Y(t) = (X(t), X(t-1), \ldots X(t-n+1)).$$

This possibility is useful in models of learning (see Chapter 9). Although not conceptually difficult, this operation may sometimes lead to computational difficulties. For example, if $n = 3$ and $X(t)$ can take say 100 values, then the storage of $Y(t)$ requires $(100)^3 = 10^6$ values. This is a larger memory requirement than the capability of many microcomputers. This illustrates a more serious limitation. The solutions of dynamic optimization problems rapidly increase in difficulty (whether treated numerically or analytically) as the dimension of the state space increases. This, however, is not only a limitation of the approach used here, but is an inherent feature of all dynamic optimization problems (Bellman 1957). We discuss these limitations in more detail in the next chapter.

The reasons underlying our preference for a parsimonious selection of state variables should now be clear. Even though a biologically accurate description of any organism may require a large number of state variables (McFarland and Houston 1981), the twin difficulties of estimating a large number of physiological and behavioral parameters, and of solving a high-dimensional dynamic optimization problem, force one to make simplifying assumptions, and to retain a small number of state variables. However, as we have repeatedly shown, even the introduction of one or two state variables can often lead to important new insights into the adaptive significance of observed behavioral patterns.

How restrictive is the assumption of a discrete time parameter? Note that we only allow a single state transition to occur in each time period $t$ (see Figure 8.1). Also, only a single behavioral decision occurs in each period. The modeler must therefore devote some care to the selection of the basic time unit. If a bird switches patches on the average once every five minutes, then a five-minute basic time interval may be appropriate for modeling patch selection. But if we are modeling residence-time strategy, a much shorter basic time unit, say 10 s, might be required. On the other hand, a model of foraging and migration might well employ a basic time unit of one day.

In fact, however, the situation is more flexible than it may first appear. Dynamic submodels can be linked sequentially (e.g. models of breeding and nonbreeding seasons, in alternating sequence), and short-term models can be used as submodels of long-term models (e.g. a model of behavior over a single day could be used as a submodel of behavior over a season). The method of sequential coupling (see Appendix 2.1.5) facilitates the use of such submodeling techniques.

Discretization is a normal procedure throughout computational mathematics; for example, continuous-time problems usually must be discretized for the purpose of numerical computation. It is our belief that formulating behavioral models in discrete time *ab initio* will seldom lead to poor results, provided that the basic time unit is chosen with care. Even in cases where state transitions occur at random times, it is usually possible to employ a discrete-time framework by using a sufficiently short basic time unit.

## 8.2 The Optimization Criterion: Fitness

The principle of evolution by natural selection suggests that the appropriate optimization criterion for behavioral models is lifetime fitness. In this book we have not provided a "grand" or "unifying" definition of fitness, but instead have shown how the definition of fitness arises in a natural way in each instance, as a consequence of the basic biology of the problem. Typically, we have adopted a definition of lifetime fitness in terms of survival and reproduction. We recognize that this definition may require qualification because of complications that we've not addressed. For example:

(a) Fecundity of offspring may depend on their size, which is affected by parental behavior. In such cases, fitness is better defined in terms of the number of viable grandchildren (see Chapter 4).

(b) The *timing* of reproduction may affect fitness.

(c) Behavior may affect the fitness of kin. Here the appropriate notion is called ***inclusive fitness*** (Hamilton 1964, Dawkins 1976, Grafen 1984, Brown 1988).

(d) In environments with large-scale fluctuations, the expectation of the logarithm of total reproduction (instead of reproduction) may be a more appropriate definition of fitness. In the appendix to this chapter, we show how this definition fits readily into the dynamic, state variable framework.

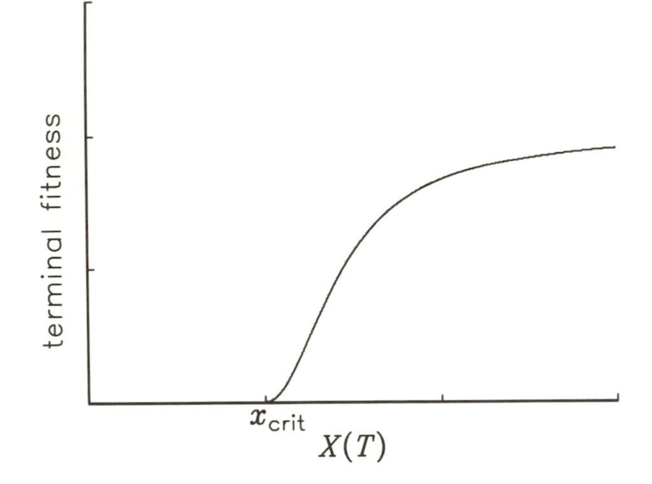


Figure 8.2 A typical terminal fitness function $\phi(X(T))$

TERMINAL FITNESS FUNCTION

The simplest case arises when behavior is only being treated over a nonreproductive period $t = 1, 2, \ldots, T$. In this case, the animal's terminal state $X(T)$ by definition completely determines its ultimate expected reproduction.* Hence we can represent lifetime fitness as

$$\text{fitness} = E\{\phi(X(T))\}. \tag{8.8}$$

As in Chapter 2 the function $\phi$ is called the *terminal fitness function*. Typically this function will have the shape shown in Figure 8.2.

Terminal fitness might depend on more than one state variable. For example, the reproductive potential of male mountain sheep (*Ovis canadensis*) probably depends both on body size $X(T)$ and harem size $Y(T)$ at the beginning of the breeding season. A hypothetical terminal fitness function for the nonbreeding season, of the form

$$\phi(X(T), Y(T)) = \phi_1(X(T)) \cdot Y(T)$$

might thus not be unreasonable. If a harem can only be built up at the cost of reducing body weight or risking injury, then the station-

* The word "terminal" here refers only to the nonbreeding period under consideration, and not to the animal's life span.

arity argument (see Section 2.4) would lead to the prediction that harems would only be sought as the breeding season approached. Similar considerations would apply to models involving territory size, for example.

### Reproduction

Terminal fitness applies to reproduction that occurs beyond the time horizon of the model. Very often, however, we wish to include reproductive behavior within the time horizon. Let the reward function $R(X(t), A(t), t, w(t))$ (see Figure 8.1) represent offspring produced in period $t$. The objective function of the model would then become

$$E\left\{ \sum_{t=1}^{T-1} R(X(t), A(t), t, w(t)) + \phi(X(T)) \right\}. \tag{8.9}$$

This definition is clearly consistent with our concept of fitness involving survival and reproduction. Equation (8.9) is the general optimization criterion for our dynamic modeling framework. The special case where only terminal fitness counts is obtained simply by setting $R \equiv 0$.

Note that Eq. (8.9) involves the expectation operator, relative to the random variables $w(1), w(2), \ldots, w(T)$—in other words, a $T$-fold expectation. This may seem hopelessly complicated, but the method of dynamic programming allows us to proceed iteratively, so that only one expectation at a time ever needs to be calculated.

## 8.3 The Dynamic Programming Algorithm

The general form of the dynamic optimization problem is:

$$\underset{\{A(t)\}}{\text{maximize}} \quad E\left\{ \sum_{t=0}^{T-1} R(X(t), A(t), t, w(t)) + \phi(X(T)) \right\} \tag{8.10}$$

$$\text{subject to} \quad X(t+1) = G(X(t), A(t), t, w(t)). \tag{8.11}$$

The actions $A(t)$ are assumed to be selected from some given set $S$, possibly time-dependent. The state variable(s) may be subject to various constraints, but these can be included within the general notation of Eq. (8.11).

We define the *lifetime fitness function* $F(x,t,T)$ as:

$$F(x,t,T) = \text{maximum}\, E\Big\{ \sum_{j=t}^{T-1} R(X(j), A(j), j, w(j)) + \phi(X(T)) \mid X(t) = x \Big\} \qquad (8.12)$$

where the maximum refers to choices of actions $A(j)$, $j = t, t+1, \ldots, T-1$. Thus $F(x,t,T)$ is in fact the *expected future lifetime reproduction of an animal whose state at the start of period $t$ equals $x$.*

We then have, first,

$$F(x,T,T) = \phi(x). \qquad (8.13)$$

In deriving the dynamic programming equation, it is often helpful to think first about the case $t = T-1$. Suppose that a particular strategy $A$ is adopted. If $w(T-1)$ takes the value $w$, then since $X(T-1) = x$ by definition (8.11), we have

$$X(T) = G(x, A, T-1, w).$$

The optimal fitness $F$ in the final period $T$ will therefore be equal to $F(G(x,A,T-1,w),T,T)$, which is already known, by Eq. (8.13). Hence the future fitness from period $T-1$ on is equal to

$$R(x,A,T-1,w) + F(G(x,A,T-1,w),T,T).$$

In order to obtain the maximum total expected fitness from period $T-1$ on, we must simply maximize the expectation of this:

$$F(x,T-1,T) = \max_A E_w \{R(x,A,T-1,w) + F(G(x,A,T-1,w),T,T)\}. \qquad (8.14)$$

By repeating the above argument, we see that in general, for $t = 1, \ldots, T-1$

$$F(x,t,T) = \max_A E_w \{R(x,A,t,w) + F(G(x,A,t,w),t+1,T)\}. \qquad (8.15)$$

This is the general dynamic programming equation. Several comments are worth making at this point.

First, the intuitive content of the dynamic programming equation is the following. The animal must select some action $A$ in period $t$. The expected immediate reward is $E_w\{R(x, A, t, w)\}$. A "shortsighted" animal might simply maximize this immediate reward, but a "prudent" animal would consider the future as well. The optimized future fitness (*after* period $t$), is by definition $F(X(t+1), t+1, T)$ where $X(t+1) = G(x, A, t, w)$ with probability $f(w)$. Total fitness, which equals total reproductive success, is the sum of these two terms. Finally, the optimally fit animal must maximize its total lifetime fitness "from now on." This is exactly what Eq. (8.15) says.* The idea of maximizing total lifetime fitness at each period $t$ was the basis of R.A. Fisher's notion of *reproductive value* (Fisher 1930).

Second, in deriving Eq. (8.15) we assumed that the animal's choice in each period $t$ is made before stochastic events occur. This would apply, for example, to group formation in lions, as in Chapter 3. In some cases the temporal sequence may be reversed, with the stochastic events occurring before decisions are made, as in Chapter 4 where we considered oviposition decisions of insects. In such cases the form of the dynamic programming equation will be somewhat different from Eq. (8.15).

Third, the dynamic programming equation provides an operational algorithm for computing both future lifetime fitness, and the optimal strategy $A$, at each period $t$. Specifically, we solve Eq. (8.15) iteratively (backwards in time) for $t = T-1, T-2, \ldots, 1$. Each iteration provides both the value of $F(x, t, T)$ and the optimal strategy $A^* = A^*(x, t)$. Note in particular that $A^*$ depends on the current state $x$ and the current time period $t$. Such a strategy specification is called a *feedback control policy*. The fact that what an animal should do may depend on its current state, and also on the time (e.g. of day, or of year) is one of the most important features of the dynamic state variable approach to the modeling of behavior.

How is the expectation $E_w\{\ldots\}$ in Eq. (8.15) actually computed? The probability distribution for $w$ is specified as part of the model; in general it may depend on the current time $t$, the current state $X(t)$, and on the chosen action $A$:

$$\Pr(w(t) = w_{ij}) = \lambda_{ij} \qquad (j = 1, 2, \ldots, N_i) \quad \text{if } A_i \text{ is chosen}$$

* The observation that a strategy which is optimal from $t$ to $T$ must, regardless of the decision taken in period $t$, remain optimal from $t+1$ to $T$, is called Bellman's "principle of optimality."

Here $w_{ij}$ and $\lambda_{ij}$ may depend on $t$ and $X(t)$. We therefore have, for any function $Q(w)$

$$E_w\{Q(w)\} = \sum_{j=1}^{N_i} \lambda_{ij} Q(w_{ij}) \qquad \text{if } A_i \text{ is chosen.}$$

This specifies how the expectation is computed in Eq. (8.15).

### 8.3.1 Computer Realization

The writing of a computer program to solve the dynamic programming equation is greatly simplified by the fact that the basic optimization subroutine can be used repeatedly. The following "pseudocode" outlines the steps actually needed in writing such a program:

(1) Input data.
(2) Initialize $F(x,T,T)$ (see Eq. 8.13).
(3) Set $t = T - 1$.
(4) For each $x$:
    For each strategy $A$:
    Compute $V_A = E_w\{R(x,A,t,w) + F(G(x,A,t,w),t+1,T)\}$.
    Find $V_{A^*} = \max V_A = F(x,t,T)$.
    Store $F(x,t,T)$.
    Print $A^*$ and $F(x,t,T)$ if desired.
(5) Subtract 1 from $t$ to get new $t$.
(6) If $t \geq 1$ go to Step 4. Otherwise stop.

Each of these steps should be coded as a subroutine (see the Addendum to Part I).

### 8.3.2 Discretization and Interpolation

Most state variables encountered in practice are capable of taking on a continuum of values. The fitness function $F(x,t,T)$ thus generally depends upon a continuous variable $x$. Unfortunately, the computer is only capable of storing finitely many discrete values $F(x_1,t,T),\ldots,F(x_n,t,T)$. Consequently, in the process of writing your program, you will usually have to decide on some discretization of the state variable $x$. This is a matter of choice, and may require some testing in order to obtain an acceptable discretization. The fitness function $F(x,t,T)$ will normally be stored as an array in $x$. (Note that $T$ is not really a variable, and can be suppressed in the program. We maintain the notation because it greatly assists in keeping one's ideas straight.)

In most cases it will not happen that $X(t+1) = G(x, A, t, w)$ will turn out to be equal to one of the chosen discretized values for $x$. In this case, some form of approximation must be used in order to calculate $F(X(t+1), t, T)$ from the stored array. We recommend simple linear interpolation (see Figure 8.3). Pseudocode for such an interpolation routine is given below; this can be coded up as a subroutine which can be included in any program. Assume that $F(x_i, t, T)$ is stored as the array element $B(i)$ for $i = 0, 1, \ldots, n$; let $\Delta x = x_{i+1} - x_i$ denote mesh size in the assumed discretization of $x$. In order to compute the interpolated value of $F(x, t, T)$ for some $x$ with $x_0 \le x \le x_n$:

(1) Determine the largest integer $i$ such that $x_i \le x$:

$$i = \text{int}\left(\frac{x - x_0}{\Delta x}\right)$$

where int $(z)$ is the integer part of $z$.

(2) If $x = x_i$ then $F(x, t, T) = B(i)$.

(3) Otherwise set

$$F(x, t, T) = \frac{(x - x_i)B(i+1) + (x_{i+1} - x)B(i)}{x_{i+1} - x_i}. \tag{8.16}$$

Note that (8.16) is just the equation of the straight line joining the points $(x_i, B(i))$ and $(x_{i+1}, B(i+1))$—see Figure 8.3.

Interpolation can also be used for the case of multidimensional state variables $X = (X_1, \ldots, X_m)$, although the method becomes cumbersome for more than two state variables. Here we illustrate it for the case in which there are two state variables $X(t)$ and $Y(t)$, so we want to find $F(x, y, t, T)$. As before, define $x_i$ and $y_j$ respectively by the condition that $i$ and $j$ are the largest integers such that $x_i \le x$ and $y_j \le y$, and assume that $F(x_l, y_m, t, T)$ is stored in the array $B(l, m)$. Then if

$$x_i = x \quad \text{and} \quad y_j = y$$

we're finished: $F(x, y, t, T) = B(i, j)$. Otherwise, define $p$ and $q$

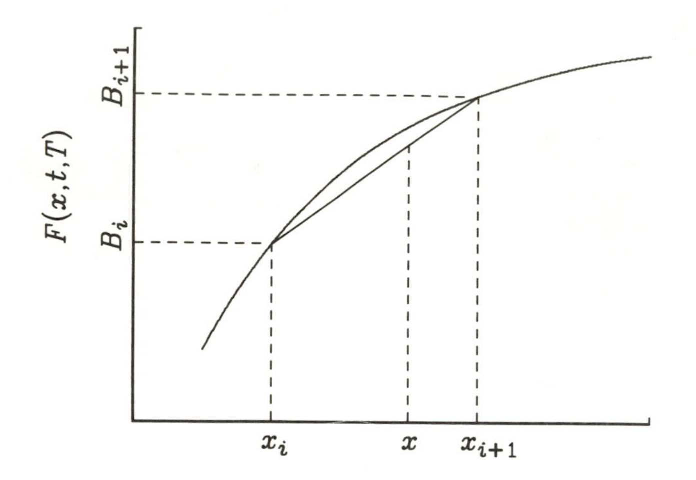


Figure 8.3 Linear interpolation of the fitness function: $F(x,t,T)$ is approximated by the line through $(x_i, B(i))$ and $(x_{i+1}, B(i+1))$, where $B(i) = F(x_i, t, T)$.

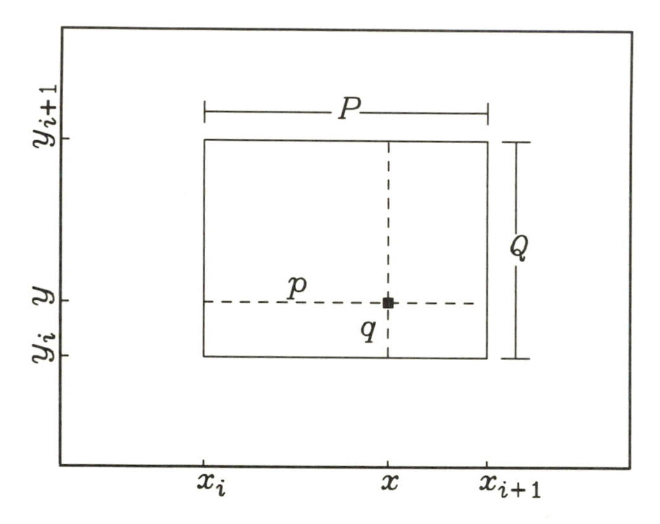


Figure 8.4 Two-dimensional interpolation

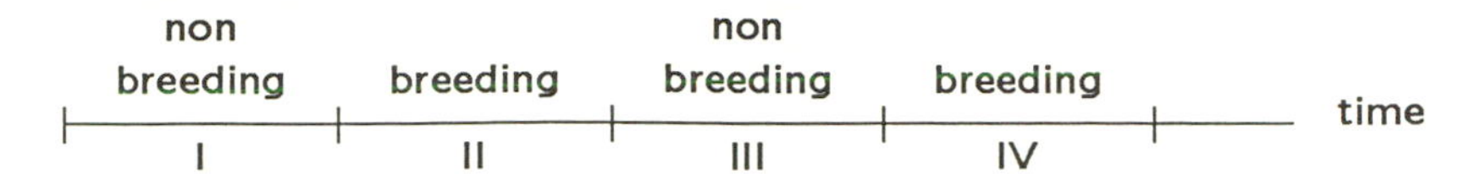


Figure 8.5 A sequence of breeding and nonbreeding seasons.

(see Figure 8.4) and $P$ and $Q$ by

$$\left.\begin{array}{ll} p = x - x_i & P = x_{i+1} - x_i \\ q = y - y_j & Q = y_{j+1} - y_j . \end{array}\right\} \tag{8.17}$$

Then a suitable interpolation formula is

$$\begin{aligned} F(x,y,t,T) = \frac{1}{PQ} & [pqB(i+1,j+1) + p(Q-q)B(i+1,j) \\ & + (P-p)qB(i,j+1) + (P-p)(Q-q)B(i,j)] . \end{aligned} \tag{8.18}$$

Eq. (8.18) can be generalized to more than two dimensions (see Abramovitz and Stegun 1965).

### 8.3.3 Sequential Coupling

By our current definition* fitness has the property of being additive over different time periods. This is immediately recognized in the dynamic programming equation (8.15) which expresses lifetime fitness $F(x,t,T)$ in terms of immediate reproduction $R(x,A,t,w)$ and future fitness $F(X(t+1),t+1,T)$. The probability of survival must also be taken into account, and our expectation operator $E_w$ automatically does this.

We can take advantage of this time additivity to simplify the computation of lifetime fitness in cases where, for example, breeding and nonbreeding seasons are interspersed, as in Figure 8.5.

A behavioral optimization model can first be constructed for each subperiod I, II, ... independently (only two models might be needed for the example of interspersed breeding and nonbreeding seasons, although parameters might change as the animal's age increased).

Let $\phi_i(X(T_i))$ be the terminal fitness function for stage $i =$ I, II, .... Notice that the future fitness function $F_i(x,1,T_i)$ at

* An alternative definition is discussed in the appendix to this chapter.

the beginning of period $i$ is equal to the terminal fitness function $\phi_{i-1}(x)$ for period $i-1$:

$$\phi_{i-1}(x) = F_i(x, 1, T_i).$$

Thus, starting with say $\phi_{IV}(x) = 0$ (no further reproduction possible after subperiod IV), we can obtain $\phi_{III}(x)$ by solving the dynamic programming equation for period IV, and so on backwards to period I. This process of ***sequential coupling*** can greatly simplify the dynamic modeling of behavior. An example was discussed in Chapter 6.

### 8.3.4 Stationarity

Using dynamic programming, we can compute the optimal action, or strategy, $A = A(x, t)$ in feedback form, as a function of both the current state variable $x$ and the current time $t$. A strategy $A = A(x)$ which depends only on $x$ and not on $t$ is called a ***stationary*** strategy.

The optimal strategy in a dynamic model is almost never stationary for all periods $t = 1, 2, \ldots, T$. However, under certain conditions it turns out that the optimal strategy is *almost* stationary when the time to go, $T - t$, is large. For $t$ near $T$ the optimal strategy may vary strongly with $t$.

In many cases the optimal strategy converges quite rapidly towards a stationary strategy, as the number of iterations in the dynamic programming solution (i.e. the time to go) increases, and you should be on the lookout for this. In order to obtain an (almost) stationary optimal strategy, we must start with a stationary model, meaning that the model parameters do not change over time. Thus the functions $R$ and $G$ in Eqs (8.10)–(8.11) must not depend explicitly on $t$, and furthermore, the random variables should be independent, and have a distribution that does not depend on $t$.*

Equally important is the fact that the stationary optimal strategy is independent of the terminal fitness function $\phi(X(T))$, at least if $\phi(X(T))$ is not identically zero. Changes in $\phi(X(T))$ will affect the optimal strategy for $t$ near $T$, but changes in $\phi(X(T))$ will have little effect on the optimal strategy for $t \ll T$.

* In other words, the random variables $w(t)$ must be independent and identically distributed (i.i.d.).

These facts are encouraging, in terms of the predictive value of dynamic modeling. We can make quantitative behavioral predictions and test them against field data, without having to worry unduly about accurately specifying the terminal fitness function. The prediction may be less reliable for times near the transition from one season to the next, unless the terminal fitness is accurately modeled.

Further discussion of stationarity and its usefulness in behavioral ecology is given by McNamara and Houston (1982).

## 8.4 Alternative Modeling Approaches

In this book we have devoted our attention almost exclusively to the state variable, stochastic dynamic programming approach to behavioral modeling. Does this mean that we advocate dynamic programming as the one and only best modeling approach in this area? Not necessarily, although we ourselves have become progressively more impressed with the scope and usefulness of this method. We've spoken with many biologists (some with a longstanding antipathy towards mathematics), working on many different organisms. The interest exhibited in the new approach to behavioral modeling has been encouraging.

Perhaps the major disadvantage of the dynamic programming approach is its failure to yield simple analytic solutions. Because analytic results can be extremely useful in developing insights and suggesting general principles, we believe that further use and development of analytic models is worthwhile. Many of the principles derived from analytic models will show up in the numerical output of dynamic models. Indeed, the latter can be thought of as tests of robustness of analytically derived predictions.

In what follows we shall describe briefly some of the better known alternative approaches to behavioral modeling, and outline their relationship to the state variable, dynamic programming approach.

### 8.4.1 Average-Rate Models

The usual "currency" used in classical foraging theory is the long-term average rate of net energy gain while foraging. Stephens and Krebs (1986) provide a detailed discussion of this criterion as well as some of its alternatives.

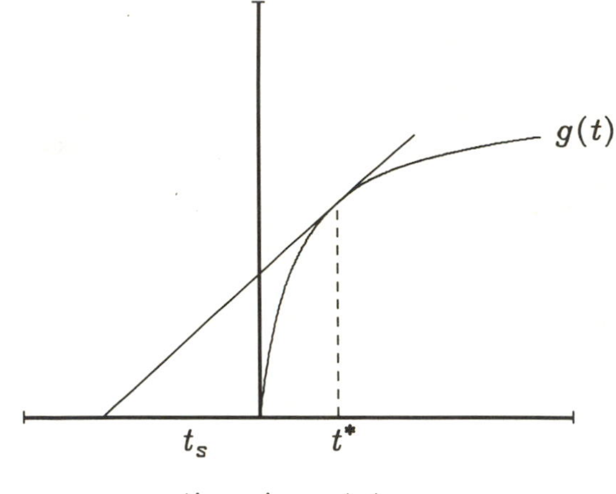


Figure 8.6 Geometrical interpretation of the marginal-value theorem: The optimal patch residence time $t^*$ equates marginal energy gain $g'(t^*)$ with the long-term average $g(t^*)/(t^* + t_s)$—see Eq. (8.20).

A well-known example of the average-rate criterion occurs in the marginal-value theorem of Charnov (1976). Consider a forager encountering equivalent patches of food by random search. The average time required to find a patch is denoted by $t_s$. The expected net energy gain from a patch is a function $g(t)$ of the time $t$ spent feeding in the patch. If time $t$ is spent in each patch, then by the renewal theorem (Appendix 1.4) the forager's long-term average rate of food intake is given by

$$R(t) = \frac{g(t)}{t + t_s}. \tag{8.19}$$

A necessary condition for $R(t)$ to be maximized is that $R'(t) = 0$. Carrying out the differentiation and simplifying, we obtain the equation

$$g'(t) = \frac{g(t)}{t + t_s} = R(t). \tag{8.20}$$

This equation asserts that, at the optimal patch-residence time $t$, the marginal rate of net energy gain $g'(t)$ equals the long-term average rate $R(t)$. This is the marginal-value theorem. If the function $g(t)$ is given, the optimal value of $t = t^*$ determined by Eq. (8.20) can be obtained graphically as shown in Figure 8.6.

This simple model is easily extended to the case of several patch types, with gain functions $g_i(t)$. The marginal-value theorem then becomes (Stephens and Krebs, 1986, p. 28)

$$g_i'(t_i) = R(t_1, t_2, \ldots, t_n) \tag{8.21}$$

where $R(t_1, t_2, \ldots, t_n)$ denotes the average rate of energy intake over all patch types. The marginal-value theorem leads to qualitative and quantitative predictions. For example, from Figure 8.6 it follows easily that foragers should spend more time in each patch when patches are hard to find (i.e. when $t_s$ is large). Also, the marginal rate of energy gain just before leaving a patch should be the same for all patches (see Eq. 8.21). Experimental and field tests of these predictions are discussed by Stephens and Krebs (1986, pp. 187ff); see also Krebs et al. (1983).

Taking temporal averages is the simplest way to attach a one-dimensional optimization objective ("currency") to a dynamic process. But this simplicity is obtained at the cost of limitations which may be severe. First, if food is patchily distributed, then variances in food intake may also be important; mean-variance models are discussed below. Second, identifying fitness with average rate of food intake either ignores survival and reproduction, or at best tacitly assumes that they are closely correlated with food intake. If this is not the case, then the forager is faced with the problem of optimizing the tradeoff between two or more competing objectives (Mesterton-Gibbons 1988). A state variable model is capable of treating such tradeoffs in a realistic manner. Third, average-rate models assume that all parameters are constant and fully known to the organism; these models cannot be used to address problems of information and learning. Dynamic state variable models are capable in principle of dealing with variable and uncertain parameters (see Section 9.1).

Average-rate models will continue to be useful as sources of insight into behavioral phenomena. However, close quantitative agreement between theory and observation (especially field observation) is not likely often to be encountered, owing to the limitations described above.

### 8.4.2 Mean-Variance Models

Various methods have been proposed for including variance in behavioral models. The $Z$-score model (Stephens 1981, Stephens and Charnov 1982), for example, replaces the objective of maximizing average rate of food intake with that of maximizing the probability of survival over a given time period. Suppose that food intake over the period is a random variable $X$, and let $F(x)$ denote the cumulative distribution function of $X$. The organism's behavior may influence both the mean and variance of $X$, i.e. $F(x)$ may depend on behavior (patch choice, etc.). Now assume that the organism survives the period if and only if food intake $X$ exceeds some minimal requirement $R$. We then have

$$\Pr(\text{organism survives period}) = \Pr(X > R) = 1 - F(R). \quad (8.22)$$

If we now assume that $X$ is normally distributed with mean $\mu$ and variance $\sigma^2$, we have

$$F(R) = \Phi(Z)$$

where

$$Z = \frac{R - \mu}{\sigma} \quad (8.23)$$

and where $\Phi$ is the normal c.d.f.

$$\Phi(z) = \frac{1}{\sqrt{2\pi}} \int_{-\infty}^{z} e^{-t^2/2}\, dt. \quad (8.24)$$

Since $\Phi(z)$ is increasing in $z$ it follows from Eq. (8.22) that maximizing the probability of survival is equivalent to minimizing the "$Z$-score" given by (8.23).

For example, consider a forager facing a choice of one of two patches $P_i$, with the same mean $\mu_1 = \mu_2$ but different variances, with $\sigma_1^2 < \sigma_2^2$. The optimal patch is that which minimizes $Z = (R - \mu)/\sigma_i$. The decision rule is therefore:

$$\left.\begin{array}{l}\text{choose patch 1 if } R < \mu \\ \text{choose patch 2 if } R > \mu\end{array}\right\} \quad (8.25)$$

In other words, the forager should be risk-averse (patch 1) if the expected food intake $\mu$ exceeds the requirement $R$, and vice versa (cf. Figure 3.2).

This simple prescription is in fact a fairly robust principle for any choice among risky projects. Roughly speaking, it says "when desperate gamble, when well off play safe." The $Z$-score model (which applies to an arbitrary number of patches $P_i$ with arbitrary means and variances) provides a simple pedagogic demonstration of this general principle. Note, however, that the $Z$-score model pertains only to a single time period. In reality, foragers must hope to survive for a sequence of time periods, and may be able to switch patches in each period. The requirement $R(t)$ in period $t$ will generally depend on the forager's past successes. Extending the $Z$-score model to a sequential decision framework automatically indicates the use of a dynamic state variable model of the type discussed in Chapter 2.

An alternative approach to risk sensitivity employs the concept of a utility function $U(x)$ (see Section 2.2). It is assumed that $U(x)$ is increasing in $x$. Decision makers are hypothesized to maximize their expected utility $E_X\{U(X)\}$. It then follows that decision makers will be risk-averse if $U$ is concave ($U'' < 0$) and risk-prone if $U$ is convex ($U'' > 0$). More generally, if $U$ is concave-convex then the general principle, "when desperate gamble, etc.," applies (see Appendix 2.2).

In order to apply utility theory to animal behavior one must first specify the utility function $U(x)$. Ideally this should be identified with fitness. Assuming that $x$ represents the (random) outcome of a single decision, we thus again face the problem of specifying a fitness currency, in this case conditional on the outcome $x$.

The $Z$-score model, for example, can be obtained as a special case of utility theory, with utility function given by

$$\begin{aligned} U(x) &= \Pr(\text{organism survives period} \mid \text{food } x \text{ is discovered}) \\ &= \begin{cases} 1 & \text{if } x \geq R \\ 0 & \text{if } x < R. \end{cases} \end{aligned} \tag{8.26}$$

We then have

$$\begin{aligned} E\{U(X)\} &= \int_{-\infty}^{\infty} U(x)\, dF(x) \\ &= \int_{R}^{\infty} dF(x) = 1 - F(R) \end{aligned} \tag{8.27}$$

which is the same as Eq. (8.22).

Utility theory, which has been the main framework used in the decision sciences, has a long history and an extensive literature discussing its applications and limitations (e.g. Machina 1987).

Most of this literature pertains to single decisions, and tends to ignore both past successes and future prospects. In a study of human decision making under risk, Kahneman and Tversky (1979) conclude that utility theory is a poor predictor of human behavior. How useful utility theory might prove for animal behavior remains to be seen. Like average-rate theories, basic utility theory has the advantages of relative simplicity and generality. Its restriction to single decisions results in limitations similar to those of average-rate models (although variance is now taken into consideration).

As pointed out in Section 2.2, in a dynamic framework the utility function $U(x)$ can be naturally identified with the lifetime fitness function $F(x,t,T)$. The dynamic programming equation then implies that organisms maximize expected utility, depending on the current state and time. We maintain that this is the ideal way to link utility theory with behavioral ecology; there is no need to import a separate notion of utility from economics into biology.

### 8.4.3 Life-History Models

Life-history models have played a central role in theoretical ecology since their introduction by R.A. Fisher (1930). In Section 2.6 we showed that life-history models are a special case of dynamic state variable models in which the state variables are suppressed, and behavior is not responsive to the current state of the organism or its environment. Life-history models have been used to study the adaptive significance of phenomena such as iteroparity vs. semelparity (Murphy 1968) and sequential hermaphroditism (Charnov 1982).

Life-history models may be called dynamic, at least in the sense that organisms' ages are explicitly taken into consideration. Both survival and reproduction are explicitly treated, and the definition of fitness is therefore directly related to evolutionary success. These models will remain useful for the analysis of life-history strategies, but their structure does not permit application to the study of behavior in the usual sense. The analysis of facultative behavior, which is responsive to internal states or external events, requires the use of dynamic state variable models.

### 8.4.4 Optimal Control Theory

Optimal control theory is a collection of mathematical results designed to lead to the solution of dynamic optimization problems.

Designed primarily for engineering applications, optimal control theory has also proved useful in the theory of economic dynamics (Kamien and Schwartz 1981). In its standard formulation, optimal control theory models system dynamics in continuous time, by means of a differential equation

$$\frac{dx}{dt} = G(x, u, t), \qquad 0 \leq t \leq T, \qquad x(0) \text{ given} \tag{8.28}$$

where $x = x(t)$ denotes the state variable and $u = u(t)$ the control or decision variable. In general both $x$ and $u$ may be multidimensional. Either or both of $x$ and $u$ may be subject to constraints.

The optimization objective is expressed as an integral:

$$J = \int_0^T L(x, u, t)\, dt + \Phi(X(T)) \tag{8.29}$$

where $L(x, u, t)$ is the payoff function and $\Phi(X(T))$ denotes a terminal reward. The mathematical problem is then to determine the control function $u(t)$ that maximizes this objective, subject to the given constraints. Note the similarity of this structure to our general dynamic optimization framework: state dynamics, strategies (controls), constraints, and an optimization criterion are all included. The formulation is deterministic rather than stochastic, however.

A collection of necessary conditions, called the Pontrjagin Maximum Principle, can sometimes be used to determine the optimal control $u^*(t)$ and optimal state dynamics $x^*(t)$. Unless the original problem is particularly simple (e.g. one-dimensional), however, the solution must be obtained by numerical computation. There are many intricate technicalities, including discontinuous switching, numerical instability, nonsufficiency of the necessary conditions, convexity assumptions, and so on. Because of these difficulties, optimal control theory has been used sparingly in behavioral ecology (see e.g. Katz 1974, Oster and Wilson 1978). Gilliam (1982) has developed an optimal-control model for the tradeoff between food intake and predation risk. The connection between optimal control theory and (continuous-time) life-history theory has been discussed by Schaffer (1983).

Stochastic continuous-time optimal control theory is discussed by Mangel (1985b), where additional references may be found. Mangel (1985b, p. 45) shows how to obtain a dynamic programming equation, in this case a nonlinear partial differential equation. Solving such an equation can be a formidable task. In this book we take the position that solving stochastic (and even, in most cases, deterministic) optimal control problems in continuous time is a job for mathematical experts. Unless they are particularly simple, these problems must be solved numerically in any case, a procedure necessitating discretization. In most cases the best approach is to discretize the problem from the outset, as advocated in this book.

To summarize, there exist several alternative approaches to the modeling of behavior in an evolutionary setting. Each approach can be valuable in its proper place. But given that most actual behavioral phenomena are complex, dynamic, and stochastic, we believe that the method described in this book will prove to be the best one in a wide variety of situations. The fact that this method encompasses such diverse approaches as life-history theory and utility theory further strengthens our argument.

## Appendix 8.1
## Fitness in Fluctuating Environments

Let us consider a population of asexually reproducing individuals living in a randomly varying environment. Let $N(t)$ denote the size of the population in year $t$, and assume that

$$N(t+1) = l(t)N(t), \quad N(0) = N_0 \text{ given} \tag{8.30}$$

where $l(t)$ is a sequence of independent, identically distributed random variables. Explicitly, suppose that

$$\Pr\left(l(t) = l_j\right) = p_j, \quad j = 1, 2, \ldots, n. \tag{8.31}$$

The averaged version of this model would be

$$N(t+1) = \bar{l}N(t) \tag{8.32}$$

where $\bar{l} = E\{l\} = \sum_j p_j l_j$ is the ordinary, or arithmetic mean of $l$. If $\bar{l} > 1$ this averaged model predicts exponential growth of the population:

$$N(t) = \bar{l}^t N_0.$$

However, as Lewontin and Cohen (1969) point out, "Even though the expectation of population size may grow infinitely large with time, the probability of extinction may approach unity, owing to the difference between the geometric and arithmetic mean growth rate."

To explain this somewhat counterintuitive result, let us transform Eq. (8.30) by using logarithms. Define

$$y(t) = \log(N(t)/N_c) \tag{8.33}$$

where $N_c$ denotes a critical extinction threshold with the property that the population goes extinct if $N(t)$ falls below $N_c$. Then Eq. (8.30) becomes

$$y(t+1) = y(t) + \log l(t) \tag{8.34}$$

from which we obtain

$$y(t) = y(0) + \sum_{s=0}^{t-1} \log l(t). \tag{8.35}$$

Dividing both sides of this equation by $t$ and letting $t \to \infty$ we obtain

$$\lim_{t\to\infty} \frac{y(t)}{t} = \lim_{t\to\infty} \frac{1}{t} \sum_{s=0}^{t-1} \log l(s). \tag{8.36}$$

Now the expression $(1/t)\sum_0^{t-1} \log l(s)$ represents the sample mean of $t$ observations of the random variable $\log l$. For large $t$ the sample mean of $\log l$ is approximately equal to the expectation of $\log l$:* Hence we have

$$\lim_{t\to\infty} \frac{y(t)}{t} = E\{\log l\}. \tag{8.37}$$

From this we can conclude that

(a) $y(t) < 0$ for large $t$, if $E\{\log l\} < 0$

(b) $y(t) \to \infty$ as $t \to \infty$, if $E\{\log l\} > 0$.

* This fact agrees with the intuitive frequency interpretation of probability (Chapter 1); mathematically it is known as the "strong law of large numbers".

Since $y(t) < 0$ means that $N(t) < N_c$ by Eq. (8.33), we have shown that the population becomes extinct with probability one, if $E\{\log l\} < 0$, but grows infinitely large as $t \to \infty$, if $E\{\log l\} > 0$.

Thus the value of $E\{\log l\}$ is critical for the long-term persistence of a population in a stochastic environment. It is possible to have $E\{\log l\} < 0$ while $\bar{l} > 1$; e.g. suppose that (Lewontin and Cohen 1966, p. 1059)

$$\Pr(l = 1.1) = 0.9$$
$$\Pr(l = 0.3) = 0.1.$$

Then a simple calculation shows that $\bar{l} = 1.02$ but $E\{\log l\} = -.035$. The population grows on average at 2% per annum, but becomes extinct with probability one.

The expression $E\{\log l\}$ can be rewritten as $\log g(l)$, where $g(l)$ denotes the geometric mean of $l$:

$$\begin{aligned} E\{\log l\} &= \textstyle\sum_j p_j \log l_j \\ &= \textstyle\sum_j \log l_j^{p_j} \\ &= \log \Pi_j l_j^{p_j} \\ &= \log g(l) \end{aligned}$$

by definition of $g(l)$ (where $\Pi_j l_j^{p_j}$ denotes the product $l_1^{p_1} \cdots l_n^{p_n}$). This explains the previous reference to the geometric mean $g(l)$ versus the arithmetic mean $\bar{l}$.

These arguments have led some authors to identify fitness in fluctuating environments with the geometric rather than the arithmetic mean of reproduction, or more precisely with $E\{\log l\}$ rather than $E\{l\}$. This turns out to be a complex topic, however, since many other aspects of population dynamics may influence the fitness concept, including age structure, density and frequency dependence, and mutation and immigration rates (see e.g. Levin et al. 1984, and Ellner 1985 for applications to seed germination strategies; also see Clark and Lamberson 1988).

In this appendix we discuss the possibility of using the modified (logarithmic) definition of fitness in a dynamic programming model. To do this, let us reconsider the patch selection model of Chapter 2, but define lifetime fitness as the expected logarithm of total reproduction. We assume that there are $p$ patches, each characterized by parameters $\alpha_i$, $\beta_i$, $\lambda_i$, $Y_i$, and also $\psi_i(x)$ = reproductive output if patch $i$ is selected and the current state is $x$. (This model is deliberately abstract, but it might be thought of as a combined model of foraging and egg laying by an insect.) For example, the first $p-1$ patches may contain food but no oviposition sites: $\psi_i(x) = 0$ for $i < p$, while patch $p$ has $\psi_p(x) > 0$ but $Y_p = 0$.

Lifetime fitness was defined in Chapter 2 as

$$F(x,t,T) = \max E\Big\{ \sum_{s=t}^{T-1} \psi_i(X(s)) \mid X(t) = x \Big\}, \tag{8.38}$$

so that fitness equals total expected reproduction. Here we consider replacing the right hand side of Eq. (8.38) by the expression

$$\max E\Big\{ \log \Big( \sum_{s=t}^{T-1} \psi_i(X(s)) \Big) \mid X(t) = x \Big\}, \tag{8.39}$$

so that fitness equals the expectation of the logarithm of total reproduction.

Recall that the dynamic programming equation for $F(x,t,T)$ as given in Eq. (8.38) is

$$\begin{aligned} F(x,t,T) = \max_i \{ &\psi_i(x) + (1-\beta_i)[\lambda_i F(x_i', t+1, T) \\ &+ (1-\lambda_i) F(x_i'', t+1, T)]\} \end{aligned} \tag{8.40}$$

where $x_i' = \text{chop}(x - \alpha_i + Y_i; 0, C)$ and $x_i'' = \text{chop}(x - \alpha_i; 0, C)$. The fitness function $F(x,t,T)$ defined by using (8.39), however, does not satisfy any dynamic programming equation. Instead, we introduce an additional state variable

$$Z(t) = \text{total reproduction from period 1 to } t-1 \tag{8.41}$$

and replace the lifetime fitness function $F(x,t,T)$ by

$$F(x,z,t,T) = \max E\{\log Z(T) \mid X(t) = x, Z(t) = z\}. \tag{8.42}$$

When $t = 1$ and $z = 0$, $F(x,z,t,T)$ is the same as expression (8.39). The end condition for $F(x,z,t,T)$ is

$$F(x,z,T,T) = \log z. \tag{8.43}$$

Also, if the organism dies ($x = 0$) with $Z(t) = z$ then its fitness is $\log z$.

$$F(0, z, t, T) = \log z. \tag{8.44}$$

The dynamic programming equation becomes, for $x > 0$

$$F(x, z, t, T) = \max_i (1 - \beta_i)\{\lambda_i F(x_i', z + \psi_i(x), t+1, T) + (1 - \lambda_i) F(x_i'', z + \psi_i(x), t+1, T)\} \tag{8.45}$$

where $x_i'$ and $x_i''$ are defined as before. Here we assume that reproduction occurs in patch $i$ whether or not food is found there. For the case where only patch $p$ contains reproduction sites, Eq. (8.45) takes the form

$$\begin{aligned} F(x, z, t, T) = \max \Big( & \max_{1 \le i \le p-1} (1 - \beta_i) \times \\ & [\lambda_i F(x_i', z, t+1, T) + (1 - \lambda_i) F(x_i'', z, t+1, T)], \\ & (1 - \beta_p) F(x_p'', z + \psi_p(x), t+1, T) \Big). \end{aligned} \tag{8.46}$$

Numerical results obtained from this model are shown in Table 8.1, (a) for the case in which fitness is defined as the expectation of the logarithm of reproduction, and (b) for the case in which fitness is simply expected reproduction. There are three forage patches, with the same parameters as the example given in Section 2.4, plus one "oviposition" patch. Each visit to the oviposition patch produces one egg, at a cost of two units of energy reserves: $\alpha_4 = 2$; also $\beta_4 = 0$. The horizon is $T = 10$, which is long enough for a stationary strategy to be reached, at $t = 1$.

Note first that the optimal patch strategies in cases (a) and (b) are almost identical, differing only for $x = 1$. In the logarithmic case, when $x = 1$ the optimal action is to oviposit in patch 4, and therefore to die immediately (since $X(2) = x - \alpha_4 < 0$). This gives the value for fitness equal to $\log 1 = 0$. Any other strategy has a positive probability that the organism will die immediately without leaving any offspring; since $\log 0 = -\infty$, this possibility must be avoided at all cost.* In the non-logarithmic case (b), on

* The computer cannot of course deal with $\log 0$. We arbitrarily replaced $\log 0$ by $-10$.

Table 8.1
Optimal patch choice, $n^*$ ($n^* = 4$ is oviposition patch) and lifetime fitness $F(x, 0, 1, T)$, with fitness defined as (a) maximum expected logarithm of total reproduction, and (b) maximum expected reproduction. (For ease of comparison the figures in (b) are the logarithms of maximum expected reproduction.) Parameter values as described in text; $T = 10$.

| Energy Reserves $x$ | (a) $n^*$ | (a) $F(x, 0, 1, T)$ | (b) $n^*$ | (b) $\log F(x, 1, T)$ |
|---|---|---|---|---|
| 1 | 4 | 0.00 | 3 | 0.24 |
| 2 | 3 | 0.49 | 3 | 0.59 |
| 3 | 3 | 0.69 | 3 | 0.74 |
| 4 | 2 | 0.81 | 2 | 0.88 |
| 5 | 2 | 0.92 | 2 | 0.99 |
| 6 | 2 | 1.03 | 2 | 1.09 |
| 7 | 2 | 1.13 | 2 | 1.18 |
| 8 | 2 | 1.22 | 2 | 1.27 |
| 9 | 4 | 1.31 | 4 | 1.35 |
| 10 | 4 | 1.40 | 4 | 1.43 |

the other hand, this desperate strategy is not optimal—instead, the insect should attempt to build up its energy reserves before ovipositing. The fitness values in column (b) always dominate those in column (a). The difference between these values indicates the degree of risk aversion introduced by taking the expectation of the logarithm of reproduction.

This model illustrates an important aspect of reproductive strategy in a stochastic environment, where extinction of an isolated population may result from a few successive bad years. In such situations natural selection should favor organisms that adopt conservative reproductive strategies which maximize the probability of yielding *some* offspring, rather than more risky strategies that would maximize expected reproduction. However, if the dynamics of the population are density dependent, then producing many offspring may be important competitively. Clark and Lamberson (1988) show, by means of Monte Carlo simulations, that for a certain class of models, the type that maximizes $E\{\log l\}$ (i.e. that maximizes the geometric mean of $l$) usually wins the evolutionary survival game. In many cases this optimal reproductive strategy is a polymorphic strategy, in the sense that a type using a mixture of conservative and risky strategies outperforms types using either pure strategy. Polymorphic diapause strategies are used by insects (Dingle 1984) and by copepods (Hairston and Munns 1984), both of which face uncertain environments. Dormancy and dispersal of seeds can also be thought of as polymorphic reproductive strategies, whereby plants avoid risks of local, short-term environmental catastrophes (Levin et al. 1984, Ellner 1985).

# 9 Some Extensions of the Dynamic Modeling Approach

In this chapter we discuss two topics that go somewhat beyond the main thrust of this book. In Section 9.1 we investigate the possibility of including learning (in the sense of estimating unknown environmental parameters) in a dynamic behavioral model. This is not particularly difficult to do in principle, but in practice it leads to drastic increases in computational complexity, rapidly surpassing the power of the largest of today's computers. Nevertheless there are at least qualitative insights to be obtained from the dynamic approach to learning.

In Section 9.2 we consider the very difficult problem of combining dynamic models with game-theoretic, or so-called ESS models of behavior. Almost no work has been done in this area, because of the severe technical (including computational) difficulties involved. But evolution is a dynamic process, and the fitness of individual organisms usually depends on the behavior of other organisms. Progress in the dynamic theory of behavioral evolutionary games would be most desirable.

## 9.1 Learning

In the main parts of this book we have assumed that all relevant parameters are known to the organism. This assumption is surely false in many if not all biological situations, but was adopted in order to simplify the dynamic modeling. In this section we discuss the problem of including learning in dynamic models.* We are concerned with how an organism assesses the conditions of its environment.

Including learning makes the analysis more realistic biologically, but it also makes the programming problems much more difficult. For this reason we will consider only simple examples in this sec-

* Stephens and Krebs 1986, Chapter 4, discuss various aspects of learning relevant to foraging theory.

tion. We hope that these examples will illustrate both the principles and the difficulties involved in modeling learning dynamically.

To begin, consider the problem of assessing a site. For example, the spiders discussed in Chapter 7 arrive at a site and must somehow determine the prey arrival rate. How can this be done? Assume that prey arrive at the site randomly so that the number of prey arriving in an interval $[0, t]$ will be Poisson distributed with parameter $\lambda$, which we presume is unknown to the spider. The informational problem is then to estimate the value of $\lambda$. (For clumped prey arrivals, the arrival distribution might have some other form, such as a negative binomial distribution with parameters $m$ and $k$, in which case the informational problem would be to estimate the values of $m$ and $k$.)

Suppose that $n$ prey arrive at the site during a time interval $[0, s]$. Then the average rate of arrival over this interval, namely

$$\hat{\lambda} = \frac{n}{s} \tag{9.1}$$

is an obvious choice as an estimate of the "true" value of $\lambda$. Standard statistics texts discuss such questions as whether this estimate is unbiased (it is*), and how much confidence can be placed in the estimate.

For behavioral theory the type of question that might be raised is, how long should the forager wait before deciding to try for a better site? Such questions turn out to be quite complicated mathematically (Oaten 1977). The forager faces the problem not only of estimating $\lambda$ for a given site, but also of estimating the distribution of arrival rates over the set of all available sites.

Let us suppose, however, that the distribution of $\lambda$ is "known" to the forager—perhaps this is part of its inherited experience. This means that a probability distribution $f(\lambda)$ is given, where

prior

$$f(\lambda)\,d\lambda = \Pr(\lambda \le \tilde{\lambda} \le \lambda + d\lambda) \tag{9.2}$$

with $\tilde{\lambda}$ the actual value of $\lambda$ at a particular site. Typically $f(\lambda)$ will have a form like that shown in Figure 9.1. The average arrival rate over all sites is then $\bar{\lambda} = E\{\lambda\} = \int \lambda f(\lambda)\,d\lambda$, and the variance is $\sigma^2 = E\{\lambda^2\} - \bar{\lambda}^2$.

* The proof that (9.1) is unbiased is extremely simple: $E\{\hat{\lambda}\} = E\{n\}/s = \lambda s/s = \lambda$.

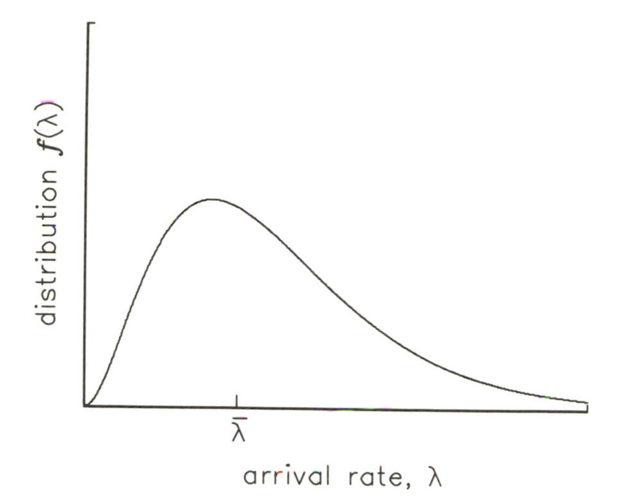


Figure 9.1 A "prior" probability distribution for prey arrival rate.

Now suppose that the forager visits a certain site for time $s$, encountering $n$ items. For example, suppose $\bar{\lambda} = 10$ items/hour, $\sigma = 5$ items/hour, $s = 1$ hour, and $n = 8$ items. Thus $\hat{\lambda} = 8$. Should the forager look for a new site? The simple-minded answer for this example is "yes"—the estimated arrival rate is below average, so the forager should move. However, there are several other matters that ought to be considered:

(1) is this really a bad site, or was the first hour's sample just unlucky?
(2) what is the cost of moving to a new site?
(3) is the present site good enough, in terms of fitness?

Let us consider the first question. A limitation of the simple estimate $\hat{\lambda} = n/s$ is that it incorporates no previous knowledge about the environment. Several methods have been devised for combining prior information with current data; we will discuss the Bayesian approach, which is based on Bayes' formula for inverse probability:

$$\Pr(A|B) = \frac{\Pr(B|A)\Pr(A)}{\Pr(B)}. \tag{9.3}$$

Bayes' formula follows immediately from the definition of conditional probability: see the proof given following Eq. (2.48).

To simplify the analysis, first assume that the unknown average prey arrival rate $\tilde{\lambda}$ at a given site can only take on one of a finite number of values $\lambda_i$, so that the prior knowledge about $\tilde{\lambda}$ is characterized by the probabilities

$$p_i = \Pr(\tilde{\lambda} = \lambda_i). \tag{9.4}$$

These are called the ***prior probabilities*** associated with the unknown parameter $\tilde{\lambda}$. In practice these probabilities would be estimated from past experience.

Since prey arrivals are assumed to follow the Poisson process, we have by Eq. (1.28)

$$\begin{aligned} P(n, \lambda, s) &= \Pr(n \text{ arrivals in } [0, s] \mid \tilde{\lambda} = \lambda) \\ &= \frac{(\lambda s)^n}{n!} e^{-\lambda s}. \end{aligned} \tag{9.5}$$

In Eq. (9.3) we now let

$A$ be the event "$\tilde{\lambda} = \lambda_i$",
$B$ be the event "$n$ prey arrivals in $[0, s]$".

The expressions on the right side of (9.3) are:

$$\begin{aligned} \Pr(B|A) &= P(n, \lambda_i, s) \\ \Pr(A) &= p_i \\ \Pr(B) &= \Sigma_j \Pr(B \mid \tilde{\lambda} = \lambda_j) \Pr(\tilde{\lambda} = \lambda_j) \qquad \text{by (1.11)} \\ &= \Sigma_j P(n, \lambda_j, s) p_j. \end{aligned}$$

Hence Eq. (9.3) becomes

$$\begin{aligned} &\Pr(\tilde{\lambda} = \lambda_i \mid n \text{ arrivals in } [0, s]) \\ &\quad = \frac{P(n, \lambda_i, s) p_i}{\sum_j P(n, \lambda_j, s) p_j} \end{aligned} \tag{9.6}$$

and this is our desired formula for estimating the arrival rate in terms of the sample data $n, s$ and the prior data $\{\lambda_i, p_i\}$. Note that we have obtained more than a point estimate for $\tilde{\lambda}$, namely we

Table 9.1
Bayesian updating of a Poisson parameter. Prior probabilities are all $p_i = 1/4$, for $\lambda_i = 1, 2, 3$, or 4. Prior mean is 2.5.

| Number of Arrivals in 1 Time Unit | Posterior Probabilities $p_1$ | $p_2$ | $p_3$ | $p_4$ | Posterior Mean |
|---|---|---|---|---|---|
| 0 | 0.64 | 0.24 | 0.08 | 0.04 | 1.52 |
| 1 | 0.43 | 0.31 | 0.17 | 0.09 | 1.92 |
| 2 | 0.22 | 0.33 | 0.27 | 0.18 | 2.41 |
| 3 | 0.09 | 0.27 | 0.34 | 0.30 | 2.89 |
| 4 | 0.03 | 0.19 | 0.36 | 0.42 | 3.17 |
| 5 | 0.01 | 0.12 | 0.34 | 0.53 | 3.35 |
| 6 | 0.00* | 0.07 | 0.30 | 0.62 | 3.52 |
| 7 | 0.00* | 0.04 | 0.26 | 0.70 | 3.66 |
| 8 | 0.00* | 0.02 | 0.21 | 0.77 | 3.75 |
| 9 | 0.00* | 0.01 | 0.17 | 0.82 | 3.81 |
| 10 | 0.00* | 0.00* | 0.13 | 0.86 | 3.83 |

*Here a value of 0.00 denotes a value that is less than $10^{-2}$.

have a (conditioned) probability distribution for the values that $\tilde{\lambda}$ can assume.

The probabilities given by Eq. (9.6) are called the *posterior probabilities* for the unknown parameter $\tilde{\lambda}$. The process of going from the prior to the posterior probabilities is sometimes called (Bayesian) *updating* of the probabilities.

If this method is new to you, here is an exercise that may help in understanding it. Assume that there are only two possible values of $\tilde{\lambda}$, say $\lambda_1 = 1$ and $\lambda_2 = 4$, with $p_1 = p_2 = 0.5$. How do prey arrivals of 0, 1, 2, 4, or 8 items in a unit time interval affect the posterior probabilities? Also find the prior and posterior expected arrival rates in each case.

A more complicated numerical example is the following. Suppose that there are four values of $\tilde{\lambda} = 1, 2, 3$, or 4. Table 9.1 shows the posterior probabilities for different numbers of arrivals. Notice how the additional data (column 1) "tightens up" the prior probabilities, so that the posterior probabilities are more heavily weighted towards the posterior mean. In particular, the variance of the posterior distribution is less than the prior variance: the degree of uncertainty about $\tilde{\lambda}$ is reduced as a result of the sampling. The Bayesian method provides a consistent and logical method for combining prior information with currently obtained sample data.

Textbooks on Bayesian analysis (e.g. de Groot 1970, Martz and Waller 1982, or Berger 1982) typically replace the discrete distribution of Eq. (9.4) by a continuous probability density for the unknown parameter, as in Eq. (9.2). Much of the science and art of Bayesian analysis is concerned with ways of choosing appropriate prior densities $f(\lambda)$. For the case of Poisson arrivals, a convenient form for the prior density is the gamma distribution

$$f(\lambda) = \gamma(\lambda; \nu, \alpha) = \frac{\alpha^\nu}{\Gamma(\nu)} \lambda^{\nu-1} e^{-\alpha\lambda} \quad (\lambda > 0). \tag{9.7}$$

This density was discussed in Appendix 1.3 of Chapter 1; the graph of $f(\lambda)$ for $\nu > 1$ is as shown in Figure 9.1.

The Bayesian updating formula corresponding to the gamma prior density can be obtained as follows. Let $f(\lambda \mid n, s)$ denote the posterior density, i.e.

$$f(\lambda \mid n, s)\, d\lambda = \Pr(\lambda \leq \tilde{\lambda} \leq \lambda + d\lambda \mid n \text{ arrivals in } [0, s]).$$

Then by Bayes' formula (9.3) we have

$$f(\lambda \mid n, s) = \frac{\Pr(n \text{ arrivals in } [0, s] \mid \tilde{\lambda} = \lambda) f(\lambda)}{\int \Pr(n \text{ arrivals in } [0, s] \mid \tilde{\lambda} = \mu) f(\mu)\, d\mu} \tag{9.8}$$

where the denominator is obtained by applying the Law of Total Probability (continuous version) to the expression $\Pr(n$ arrivals in $[0, s])$. Substituting into Eq. (9.8) from Eqs. (9.5) and (9.7), we obtain

$$\begin{aligned} f(\lambda \mid n, s) &= \frac{\frac{(\lambda s)^n}{n!} e^{-\lambda s} \frac{\alpha^\nu}{\Gamma(\nu)} \lambda^{\nu-1} e^{-\alpha\lambda}}{\int \cdots d\mu} \\ &= \text{some constant} \times \lambda^{n+\nu-1} e^{-(\alpha+s)\lambda} \end{aligned} \tag{9.9}$$

where the constant does not involve $\lambda$. Now, as remarked in Chapter 1, a conditional distribution is in particular always a bona fide probability distribution in its own right. Equation (9.9) says that

$f(\lambda \mid n, s)$ is proportional to $\lambda^{n+\nu-1}e^{-(\alpha+s)\lambda}$, which itself is proportional to $\gamma(\lambda; n+\nu, \alpha+s)$—see Eq. (9.7). Since the latter is exactly a probability density, we have to conclude without further calculation that

$$f(\lambda \mid n, s) = \gamma(\lambda; n+\nu, \alpha+s). \tag{9.10}$$

(This is another example of Professor Feynman's method of performing computations by thinking. Of course, the same result can also be obtained by brute force computation.)

Equation (9.10) giving the posterior density $f(\lambda \mid n, s)$ is analogous to the discrete case, Eq. (9.6). Note however the simple form of (9.10): the posterior density is also a gamma density, with altered parameters $\nu' = n + \nu$, $\alpha' = \alpha + s$. The posterior mean and coefficient of variation are therefore (see Table 1.1):

$$\left.\begin{aligned} \bar{\lambda}' &= \frac{\nu'}{\alpha'} = \frac{n+\nu}{s+\alpha} \\ CV' &= \frac{1}{\sqrt{\nu'}} = \frac{1}{\sqrt{n+\nu}} \end{aligned}\right\} \tag{9.11}$$

A pair of distributions, such as the gamma and Poisson, with the property that the posterior distribution has the same functional form as the prior distribution, is called a *conjugate pair* in decision theory (de Groot 1970).

To consider an example, let the prior mean of $\tilde{\lambda}$ be 5 items per hour with a $CV$ of 2. Then we have

$$\frac{\nu}{\alpha} = 5 \quad \text{and} \quad \frac{1}{\sqrt{\nu}} = 2$$

or $\nu = 0.25$, $\alpha = 0.05$. Table 9.2 shows the posterior mean and $CV$ for a range of arrival samples ($s = 1\,\text{hr}$). As one expects, the coefficient of variation $CV'$ decreases as the size of the sample increases.

The Bayesian method of updating prior information is consistent, in the sense that the posterior distribution can be treated as a prior distribution for further updating, and this produces the same result as a single updating using all the data at once. In other words, updating can be performed dynamically. As the sample size increases, the importance of the prior density is reduced, and the posterior estimates eventually depend only on the

Table 9.2
Posterior mean and coefficient of variation for the case of a gamma prior density, $\mu = 5$, $CV = 2$, $s = 1$

| Number of Arrivals $n$ | Posterior Mean $\lambda'$ | Posterior Coeff. of Variation $CV'$ |
|---|---|---|
| 0 | 0.24 | 2.00 |
| 2 | 2.14 | 0.67 |
| 4 | 4.05 | 0.49 |
| 6 | 5.95 | 0.40 |
| 8 | 7.86 | 0.35 |
| 10 | 9.76 | 0.31 |

sample data. For example, note that in Eq. (9.11) $CV' \to 0$ as $n \to \infty$, so that the posterior distribution becomes concentrated around $\bar{\lambda}'$ for large $n$ .

A disadvantage of the Bayesian approach as described here is that one must assume that the unknown parameter does not change over time. Thus neither the movement nor the depletion of a prey population can be taken into consideration (but see Mangel and Clark 1983). Methods of estimating time-varying parameters are called *filtering*. We refer the reader to texts on time series analysis for this topic (see also Stephens 1987).

## Patch Selection

Let us now consider how learning could be modeled in the simplest patch selection problem. Recall that the dynamic programming equation for that problem is

$$F(x,t,T) = \max_i (1 - \beta_i)\{\lambda_i F(x_i', t+1, T) + (1 - \lambda_i) F(x_i'', t+1, T)\} \tag{9.12}$$

where $x_i' = \text{chop}(x - \alpha_i + Y_i; 0, C)$ and $x_i'' = \text{chop}(x - \alpha_i; 0, C)$.

In this equation it is assumed that the patch parameters ($\beta_i$, $\alpha_i$, $Y_i$, $\lambda_i$) are known. Now let us suppose, however, that the probability $\lambda_i$ of encountering an item in patch $i$ is not known to the organism. If the organism makes $N_i$ visits to patch $i$ and encounters a prey item in $n_i$ of those visits, then the simple (non-Bayesian) estimate for $\lambda_i$ is

$$\hat{\lambda}_i = n_i/N_i. \tag{9.13}$$

This equation suggests that in addition to the state variable, we introduce ***informational variables*** $\mathbf{N}(t) = (N_1(t), \ldots N_p(t))$ and $\mathbf{n}(t) = (n_1(t), \ldots n_p(t))$ where $p$ is the number of different patches, $N_i(t)$ is the number of visits to patch $i$ through the end of period $t - 1$, and $n_i(t)$ is the number of times that a visit to patch $i$ yielded a prey item. Instead of $F(x, t, T)$ we must consider a new fitness function defined by

$$F(x, \mathbf{N}, \mathbf{n}, t, T) = \text{maximum expected lifetime fitness from period } t \text{ to period } T\text{, given that } X(t) = x,\ \mathbf{N}(t) = \mathbf{N} \text{ and } \mathbf{n}(t) = \mathbf{n}. \tag{9.14}$$

The end condition that $F(x, \mathbf{N}, \mathbf{n}, t, T)$ satisfies for $t = T$ will be the same as before:

$$F(x, \mathbf{N}, \mathbf{n}, t, T) = \phi(x). \tag{9.15}$$

The dynamic programming equation will be different however, since each visit to a particular patch changes both the physiological state variable and the informational state variable. In particular, if $\mathbf{N}(t) = (N_1(t), \ldots, N_p(t))$ and $\mathbf{n}(t) = (n_1(t), \ldots, n_p(t))$ and if patch $j$ is visited, then the new $\mathbf{N}$ vector is $(N_1(t), \ldots, N_{j-1}(t), N_j(t) + 1, \ldots, N_p(t))$. Denote this vector by $\mathbf{N}'_j$. If no prey item is encountered during the visit, then $\mathbf{n}(t + 1) = \mathbf{n}(t)$, while if a prey item is encountered, the new $\mathbf{n}$ vector is $(n_1(t), \ldots, n_{j-1}(t), n_j(t) + 1, \ldots, n_p(t))$. Denote this vector by $\mathbf{n}'_j$. If we use the estimate (9.13) for $\lambda_i$, the dynamic programming equation becomes

$$F(x, \mathbf{N}, \mathbf{n}, t, T) = \max_i (1 - \beta_i)\{(n_i/N_i) F(x'_i, \mathbf{N}'_i, \mathbf{n}'_i, t + 1, T) + ((N_i - n_i)/N_i) F(x''_i, \mathbf{N}'_i, \mathbf{n}_i, t + 1, T)\}. \tag{9.16}$$

Although Eq. (9.16) is fairly simple mathematically, the computer solution is somewhat unwieldy. Besides a vector of dimension $C$ for the state variable $X(t)$, the computer code must contain $2p$ vectors of dimension $T$ for the state variables $N_i(t)$, $n_i(t)$, $i = 1, 2, \ldots, p$. Each iteration requires $2pTC$ solutions of the dynamic programming equation (9.16), instead of the $C$ solutions required for the earlier case, Eq. (9.12). For example, if $p = 3$ and $T = 30$, the new equation would require at least 180 times as much computer time as the old equation. The solution will be slow on a microcomputer, but feasible on a larger machine.

However, we will now argue that Eq. (9.16) does not fully represent the informational problem. To see this simply, imagine that $N_i(1) = 0$ for some value or values of $i$—the $i$th patch has never been sampled. Then the factor $n_i/N_i$ in Eq. (9.16) is undefined. In other words, Eq. (9.16) can only be used if all patches have already been visited at least once.

Moreover, suppose now that $N_i(1) \geq 1$ for all patches, but suppose that

$$N_i(1) = 1 \qquad n_i(1) = 0$$

for some patch $i$. Then the estimate of $\lambda_i$ is $\lambda_i = 0$, so that the forager will *never* visit patch $i$—on the basis of a single failure to find prey there!

The way around these difficulties is to use a Bayesian approach, assuming a prior distribution for each $\lambda_i$, and updating that distribution after each patch visit. For the patch selection model, $\lambda_i$ is a probability, so that $0 \leq \lambda_i \leq 1$. The forager does not know the actual value of $\lambda_i$ but has a prior distribution for its value. An appropriate density in this case is the beta density, defined by

$$\beta(\lambda; \delta, w) = \frac{\Gamma(\delta + w)}{\Gamma(\delta)\Gamma(w)} \lambda^{\delta - 1}(1 - \lambda)^{w-1} \tag{9.17}$$

where $\delta$, $w$ denote positive parameters. The factor involving $\Gamma$-functions ensures that

$$\int_0^1 \beta(\lambda; \delta, w)\, d\lambda = 1.$$

It is not difficult to verify that the mean and variance of the $\beta$-distribution are given by

$$\mu = \frac{\delta}{\delta + w} \tag{9.18}$$

$$\sigma^2 = \frac{\delta w}{(\delta + w)^2(\delta + w + 1)}. \tag{9.19}$$

This two-parameter distribution is capable of representing a wide variety of distributions on the interval $0 \leq \lambda \leq 1$.

Next consider the updating formula. The forager visits patch $i$, for which his prior parameter values are $\delta, w$. Food is either discovered, or not. In the event that food is discovered, the posterior distribution for $\lambda_i$ becomes

$$\begin{aligned}
&\Pr(\lambda_i = \lambda \mid \text{food discovered}) \\
&\quad = \frac{\Pr(\text{food discovered} \mid \lambda_i = \lambda)\Pr(\lambda_i = \lambda)}{\int_0^1 \Pr \text{ food discovered} \mid \lambda_i = \mu)\Pr(\lambda_i = \mu)\, d\mu} \\
&\quad = \frac{\lambda\beta(\lambda;\delta,w)}{\int_0^1 \mu\beta(\mu;\delta,w)\, d\mu} \\
&\quad = \text{const.} \times \lambda^{\delta}(1-\lambda)^{w-1} \\
&\quad = \beta(\lambda;\delta+1,w). \qquad (9.20)
\end{aligned}$$

Here we have again used the argument that the conditional probability under consideration must be a bona fide probability, and the only density proportional to the function $\lambda^{\delta}(1-\lambda)^{w-1}$ is the density $\beta(\lambda;\delta+1,w)$.

A similar calculation shows that

$$\Pr(\lambda_i = \lambda \mid \text{food not discovered}) = \beta(\lambda;\delta,w+1). \qquad (9.21)$$

Thus we obtain a simple updating procedure, which can be summarized as follows:

| prior patch parameters | sampling result | posterior patch parameters |
|---|---|---|
| $(\delta_i, w_i)$ | find food | $(\delta_i + 1, w_i)$ |
| | don't find food | $(\delta_i, w_i + 1)$ |

From Eq. (9.18) we also see that the expected value of $\lambda_i$ increases when food is found, and decreases when it is not.

We can now derive the dynamic programming equation. Lifetime fitness $F(x, \boldsymbol{\delta}, \boldsymbol{w}, t, T)$ now depends on the informational state variables $\boldsymbol{\delta} = (\delta_1, \ldots, \delta_n)$ and $\boldsymbol{w} = (w_1, \ldots, w_n)$, which are the estimated patch parameters as of period $t$. The end condition is again simply

$$F(x, \boldsymbol{\delta}, \boldsymbol{w}, T, T) = \phi(x). \tag{9.22}$$

Now, given $\boldsymbol{\delta}$ and $\boldsymbol{w}$, the (prior) probability of finding food in patch $i$ equals $\delta_i/(\delta_i + w_i)$. Hence our dynamic programming equation becomes

$$F(x, \boldsymbol{\delta}, \boldsymbol{w}, t, T) = \max_i (1 - \beta_i) \left\{ \frac{\delta_i}{\delta_i + w_i} F(x_i', \boldsymbol{\delta}_i', \boldsymbol{w}_i, t+1, T) + \frac{w_i}{\delta_i + w_i} F(x_i'', \boldsymbol{\delta}, \boldsymbol{w}_i', t_1, T) \right\} \tag{9.23}$$

where $\boldsymbol{\delta}_i'$ and $\boldsymbol{w}_i'$ are obtained from $\boldsymbol{\delta}, \boldsymbol{w}$ by updating in patch $i$ only (since only one patch can be visited in a period).

The Bayesian model (9.23) overcomes the difficulties discussed for the non-Bayesian model (9.16). However, the computational demands in (9.23) are rather high. Suppose, for example, that the forager begins at $t = 1$ with a uniform prior distribution for $\lambda_i$ in each patch. This sets $\delta_i = w_i = 1$ initially. Either of $\delta_i$ or $w_i$ may increase by 1 each period, so that in general $\delta$ and $w$ may take the values $1, 2, \ldots, T - 1$. Since $x$ runs from 1 to $C$, the dimension of the arrays $F0$ and $F1$ used to store values of the fitness function $F(x, \boldsymbol{\delta}, \boldsymbol{w}, t, T)$ will be $(T - 1)^{2n} C$, where $n$ is the number of patches. For $n = 3$, $T = 20$, $C = 10$ this gives dimension approximately $5 \times 10^8$—far too large for microcomputers. Short of devising a more efficient algorithm (which is probably feasible), only a few iterations $T$ would be possible.

We will not pursue this example further here. Related problems discussed by Mangel and Clark (1983) and Walters (1986) show that optimal patch choice now involves not only the tradeoff between food availability and predation risk, but also the "value of information" obtained by sampling patches. Although the computations are difficult, one simple and obvious principle emerges. The forager should "probe" patches for which the prior distribution indicates large uncertainty. It may be difficult to compute the optimal amount of probing, but it is fairly easy to demonstrate that failure to probe is severely suboptimal (Mangel and Clark 1983, 1986).

## 9.2 Dynamic Behavioral Games

The models discussed thus far all treat the behavior of a single individual capable of optimizing its own fitness. A leading theme in behavioral ecology and sociobiology, however, concerns the effects of interactions between organisms. The analytical treatment of such questions relies on the theory of noncooperative games (Basar and Olsder 1983) in general, and the concept of an evolutionarily stable strategy (ESS; Maynard Smith 1982) in particular. The extensive ESS literature deals almost entirely with single behavioral decisions or sequences of identical decisions, however. Standard ESS models do not consider the current state of the animal (Zahavi 1986).

In this section, we show how—in principle—game aspects can be incorporated into the framework of dynamic, state variable modeling. We admit from the outset that the full development of a theory of dynamic evolutionary games is a formidable task, well beyond the scope of this book. For example, even the most mathematically sophisticated texts on dynamic games usually treat only the simplest of examples (e.g. pursuit–evasion games). We are not completely convinced that the engineering approach that dominates dynamic game theory at present is appropriate in biology. Even so, we can begin.

The type of game that we consider falls into the category that Maynard Smith calls "playing the field." As Houston and McNamara (1987) point out, in considering dynamic games against the field we do not have to treat complicated interactive strategies in which players continually change their behavior in response to the behavior of competitors. Thus we can restrict our attention to strategies of a fairly simple type, although this will depend on the particular situation.

In the static case (Maynard Smith 1982, p.24), the definition of an ESS in a game against the field is as follows. Let $W(I, J)$ be the fitness of a single $I$-strategist playing against a large population of $J$-strategists. A sufficient condition for $J$ to be an ESS is that

$$W(I, J) < W(J, J) \quad \text{for all } I \neq J. \tag{9.24}$$

However, it may happen that $W(I, J) = W(J, J)$ for some $I \neq J$. If this is the case, an extra condition is required, which asserts that $W(I, J) < W(J, J)$ when the population contains a small proportion of $I$-strategists. In other words, in this case a single

$I$-strategist may do as well as the $J$-strategists, but the $I$-strategy cannot spread in the population.

The above definition can be extended to the dynamic case as follows. First we define a dynamic strategy to be a sequence $I = \{i(x,t)\}$, $t = 1, 2, \ldots, T$ of actions $i(x,t)$ taken in period $t$. These actions generally depend on the organism's current state $x$. Next let $F(x,t,T;I,J)$ denote the expected lifetime fitness of a single $I$-strategist in a population of $J$-strategists. Then a sufficient condition for $J$ to be an ESS is that

$$F(x,1,T;I,J) < F(x,1,T;J,J) \text{ for all } I \neq J \text{ and for all } x. \tag{9.25}$$

If equality holds in Eq. (9.25) for the case of a single $I$-strategist, then we require strict inequality to hold when $F(x,1,T;I,J)$ denotes lifetime fitness of an $I$-strategist when these strategists constitute a small portion of the population.

## Finding the ESS—An Example

We will illustrate how the ESS can often be found in a static model by considering an example of the sex ratio game (Charnov 1982). The example will also indicate why it is likely to be much more difficult to find dynamic ESS's. In later sections we treat two examples of dynamic evolutionary games.

Consider a population containing $M$ females that reproduce simultaneously, each producing $C$ offspring in the ratio of $r^*$ males to $1 - r^*$ females. In the next generation, assume that all females are fertilized, but that the males compete for females. Now suppose that one additional female, with a different sex ratio $r$ enters the population, and let $W(r, r^*)$ denote her fitness in terms of the number of her grandchildren.

For such a mutant female, the number of grandchildren resulting from female offspring equals $(1 - r)C^2$. The total number of males in the first generation is $rC + Mr^*C$, so that the probability that a mutant male offspring fertilizes a given female is

$$\frac{rC}{rC + Mr^*C} = \frac{r}{r + Mr^*}.$$

Hence the expected number of grandchildren from male offspring is

$$\frac{r}{r + Mr^*}[(1 - r)C + M(1 - r^*)C]C.$$

This leads to the expression

$$W(r, r^*) = C^2\Big\{1 - r + \frac{r[1 - r + M(1 - r^*)]}{r + Mr^*}\Big\}. \tag{9.26}$$

Now the ESS ratio $r^*$ is characterized by the condition

$$W(r, r^*) < W(r^*, r^*) \text{ for } r \neq r^*. \tag{9.27}$$

In other words, $W(r, r^*)$ is maximized for $r = r^*$. From calculus, this implies that

$$\left.\frac{\partial W}{\partial r}\right|_{r=r^*} = 0. \tag{9.28}$$

Performing this calculation, we conclude that

$$r^* = \frac{M}{2(M+1)}, \tag{9.29}$$

a result originally obtained by Hamilton (1967) (whose result $r^* = (N-1)/2N$ is obtained by setting $M = N - 1$).

The procedure used in this example consisted of three steps: (a) formulation of a model; (b) calculation of the fitness function $W(r, r^*)$; and (c) determination of the ESS (by differentiation). For a dynamic model, step (a) can be accomplished by the procedures described in this book. In some cases steps (b) and (c) can be carried out simultaneously in principle by an iterative procedure similar to the dynamic programming algorithm. In practice, however, this computation in general becomes extremely difficult if not impossible. (An alternative algorithm is used by Houston and McNamara 1987.) We discuss two examples in the following sections.

### 9.2.1 A Dynamic Game between Tephritid Flies*

The tephritid fly *Oriella ruficada* attacks the head of the thistle plant *Cirsium arvense*. The important aspects of the natural history of the thistle and fly are the following:

(1) The thistle heads are available for only a single day. The heads vary in size and nutrient level, so that there is a probability distribution for the number of larvae (maggots) that

* We thank B. Roitberg and R. Lalonde for telling us about this problem.

a head can support. There is a maximum value $R_m$ of resource levels in a thistle head. The distribution is shown qualitatively in Figure 9.2.

(2) Flies oviposit at most 20 eggs per day. After oviposition in a thistle, a female marks the head with a pheromone. The pheromone mark probably lasts the entire day. If the number of eggs in a head exceeds the head's resource level, then the size of the flies that ultimately emerge is reduced.

The dynamic game arises in the following way. When a female encounters a clean head, how many eggs should she oviposit in the head, taking into account future ovipositions by other flies who will recognize that the fruit is marked? (We presume, to make the analysis simple, that re-encounters of a female with a fruit in which she has already oviposited do not occur.) When the mutant female encounters a marked host, her oviposition decision must take into account ovipositions both before and after her own oviposition.

In order to illustrate the dynamic game—and its complexities—we introduce several simplifying assumptions. These assumptions are used for pedagogic purposes; we admit from the outset that the model is highly unrealistic (see Table 9.3 for alternative assumptions). We assume:

(1) All thistle heads have exactly the same resource level. The fitness per egg from any head, containing a total of $c$ eggs is denoted by $f(c)$; see Figure 9.3.
(2) The time horizon is $T = 3$, so that this is a two-period problem.
(3) All flies start in period 1 with exactly the same number $X_0$ of eggs.
(4) Flies encounter exactly one thistle head per period.
(5) In period 1, all thistles are clean. In period 2, a fraction $p$ of the thistles are marked. The probability that a specified head is re-encountered in period 2 by another fly is $\rho$. The values of $p$ and $\rho$ are given exogenously.
(6) The probability that a fly survives to period 2 is $\mu$.

The effect of Assumption (3) is that the state variable enters the problem in a very simple way that allows easy formulation and relatively easy solution of this problem.

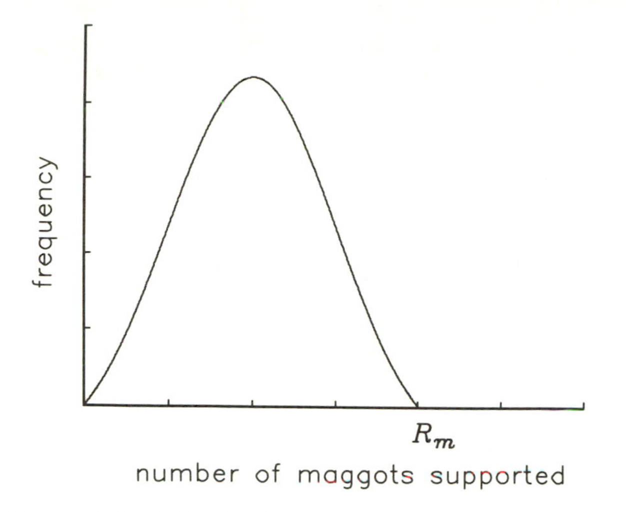

Figure 9.2 Distribution of resources in a thistle head.

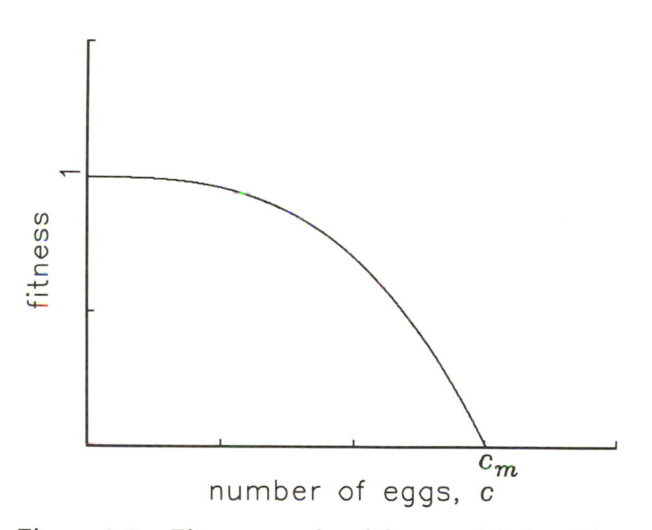

Figure 9.3 Fitness per head from a thistle with $c$ eggs.

Table 9.3
Simplifying assumptions and alternative assumptions in the thistle fly game

| | Simplifying Assumption | Alternative assumption |
|---|---|---|
| (1) | all heads have the same resource level | distribution of resources per head |
| (2) | $T = 3$ | $T$ large; probably consisting of 8 hrs of 15 minute periods ($T \approx 30$) |
| (3) | all flies start with $X_0$ eggs | distribution of eggs in flies |
| (4) | one encounter per period | multiple encounters |
| (5) | encounter parameters fixed and given exogenously | encounter rates depend on what the flies do |
| (6) | mortality risk is given exogenously | mortality risk depends on the flies' age and behavior |

A strategy $J$ for this problem consists of a triplet of decisions:

$$J = \{c_1, c_2, m_2\}$$

where

$$c_i = \text{clutch laid in a clean head in period } i \ (i = 1, 2)$$
$$m_2 = \text{clutch laid in a marked head in period 2.}$$

Note that because of Assumption (3), $c_1$ is *only* a function of $X_0$, and $c_2$, $m_2$ are only functions of $X_0 - c_1$. The state dependence of the strategy in this model is minimal.

The constraints on the decisions $c_i, m_i$ are that

$$c_1 \leq X_0$$
$$c_2 \leq X_0 - c_1$$
$$m_2 \leq X_0 - c_1.$$

We will denote the ESS by $J = \{c_1^*, c_2^*, m_2^*\}$ and the mutant strategy by $I = \{c_1, c_2, m_2\}$.

We wish to compute the lifetime fitness of the $I$-strategy against the $J$-strategy. Because of the assumptions about encounters, this computation can be done explicitly.

Consider the first period. The mutant female encounters a clean head and lays $c_1$ eggs. If this head is not re-encountered in period 2, the increment in fitness from this clutch is $c_1 f(c_1)$. If it is re-encountered, the increment in fitness from this clutch is $c_1 f(c_1 + m_2^*)$. Thus, the increment in fitness from ovipositions in the first period is

$$(1 - \rho)c_1 f(c_1) + \rho c_1 f(c_1 + m_2^*).$$

If the mutant female encounters a marked head in the second period, the increment in fitness is $m_2 f(c_1^* + m_2)$. If she encounters a clean head, the increment in fitness is $c_2 f(c_2)$. Hence, the increment in fitness from the second period is

$$pm_2 f(c_1^* + m_2) + (1 - p)c_2 f(c_2).$$

Combining these, we obtain the lifetime fitness:

$$\begin{aligned} F(X_0, 1, 3; I, J) = (1 - \rho)c_1 f(c_1) + \rho c_1 f(c_1 + m_2^*) \\ + [pm_2 f(c_1^* + m_2) + (1 - p)c_2 f(c_2)]\mu \quad (9.30) \end{aligned}$$

where $c_1, c_2, m_2$ are subject to

$$\begin{aligned} c_1 + c_2 &\leq X_0 \\ c_1 + m_2 &\leq X_0. \end{aligned}$$

Our definition of the ESS is that fitness is maximized when $I = J$. But, how do we find the ESS in actual practice? In this problem we use a method of *successive approximations*. (See Houston and McNamara 1987 for another example of this method.) The algorithm is very simple:

1. Start off with a guess for the ESS, $J = \{c_1^*, c_2^*, m_2^*\}$.
2. Maximize $F(X_0, 1, 3; I, J)$ over $\{c_1, c_2, m_2\}$. If $c_1 = c_1^*$, $c_2 = c_2^*$, and $m_1 = m_1^*$ then $J$ is the ESS.
3. Otherwise set $c_1^* = c_1$, $c_2^* = c_2$, and $m_2^* = m_2$. Return to step 2.

The algorithm works by specifying an initial guess for the ESS strategy, then solving for the values of $\{c_1, c_2, m_2\}$ that maximize the mutant's fitness, replacing the initial guess by the fitness maximizing strategy, and repeating this procedure until the two strategies are equal.

Table 9.4
Results of the method of successive approximations for a two-period problem

| Parameters | | | | | | Strategies | | Number of Iterations |
|---|---|---|---|---|---|---|---|---|
| | | | | | | Initial Guess | ESS | |
| $X_0$ | $\mu$ | $p$ | $\rho$ | $c_m$ | $c_{SHM}$ | $\{c_1, c_2, m_2\}$ | $\{c_1^*, c_2^*, m_2^*\}$ | |
| 6 | 0.8 | 0.9 | 0.1 | 7 | 5 | {2, 2, 2} | {4, 2, 2} | 1 |
| 7 | 0.8 | 0.2 | 0.1 | 7 | 5 | {2, 2, 2} | {4, 3, 2} | 3 |
| 10 | 0.8 | 0.3 | 0.2 | 7 | 5 | {3, 3, 3} | {4, 5, 2} | 4 |
| 15 | 0.8 | 0.3 | 0.2 | 7 | 5 | {5, 5, 5} | {4, 5, 2} | 4 |
| 10 | 0.8 | 0.3 | 0.2 | 10 | 7 | {3, 3, 3} | {5, 5, 3} | 3 |
| 10 | 0.8 | 0.3 | 0.2 | 14 | 9 | {3, 3, 3} | {6, 4, 4} | 1 |
| 10 | 0.8 | 0.3 | 0.2 | 4 | 3 | {3, 3, 3} | {2, 3, 1} | 4 |

We apply this algorithm to the following example. For $f(c)$ we choose

$$f(c) = 1 - (c/c_m)^4.$$

Then the fitness from a clutch $c$ that is not superparasitized is $cf(c) = c - c^5/c_m^4$, the single host maximum (SHM) clutch is

$$c_{\text{SHM}} = \left(\frac{c_m^4}{5}\right)^{1/4} = \frac{c_m}{5^{1/4}} = 0.669c_m.$$

(In general, $0.669c_m$ is not an integer, so the actual value of $c_{\text{SHM}}$ is found by comparing integer values of $c$ adjacent to $0.669c_m$.)

The results obtained from the successive approximation algorithm are shown in Table 9.4, for a variety of parameter values for $X_0$, $\mu$, $p$, $\rho$, $c_m$.

When viewing this table, remember that all flies encounter a clean head in the first period. Note that, if possible, a SHM clutch is laid in a clean head in the second period. Note also that only a few iterations are required to obtain the ESS.

### Dynamic Iteration

For the above two-period problem we were able to write down the lifetime fitness explicitly, and to determine the ESS directly using the method of successive approximations. This method is not feasible for large $T$, however. We will therefore show how to develop a dynamic iteration algorithm, similar in spirit to the dynamic programming algorithm.

The strategies depend upon $x$ and $t$. We suppose that the ESS strategy is

$$J = \left\{c_1^*(x), c_2^*(x), \ldots, c_{T-1}^*(x), m_2^*(x), \ldots, m_{T-1}^*(x)\right\}$$

and that an alternative strategy is

$$I = \{c_1(x), c_2(x), \ldots, c_{T-1}(x), m_2(x), \ldots, m_{T-1}(x)\}.$$

Also let

$$\lambda_c(t) = \Pr\text{ (encounter a clean host in period } t)$$
$$\lambda_m(t) = \Pr\text{ (encounter a marked host in period } t)$$

and let the re-encounter probability $\rho$ be defined as before.

We now define a new function $F(x, t, T; J)$ as follows:

$$F(x, t, T; J) = \text{maximum lifetime fitness of the mutant through oviposition between period } t \text{ and } T\text{, given that the normal population follows strategy } J. \quad (9.31)$$

We then have the end condition

$$F(x, T, T; J) = 0. \quad (9.32)$$

Suppose that the mutant oviposits in a clean host in period $t$. There could be up to $T - t - 1$ additional ovipositions between period $t + 1$ and $T$, but the time when they occur matters. For example, suppose three additional ovipositions occur. The number of eggs from these additional ovipositions could be any of $m_{t+1}^* + m_{t+2}^* + m_{t+3}^*$, $m_{t+1}^* + m_{t+5}^* + m_{t+9}^*$, etc. Since the number of different sequences with $j$ additional ovipositions is $\binom{T-t}{j}$, we clearly don't want to keep track of all such sequences. To avoid doing so, we make the additional simplifying assumption that

$$\left.\begin{aligned} m_t^*(x) &= m^*(x) \\ m_t(x) &= m(x) \end{aligned}\right\} \text{ for all } t. \quad (9.33)$$

It would appear, then, that the dynamic iteration could be written as

$$F(x,t,T;J) = \max_{c_t(x),m(x)} \{(1-\lambda_c(t)-\lambda_m(t))\mu F(x,t+1,T;J)$$

$$+ \lambda_c(t)\Bigg[c_t(x) \sum_{j=1}^{T-t-1} \binom{T-t-1}{j} \rho^j (1-\rho)^{T-t-1-j}$$

$$f(c_t(x)+jm^*(x)) + \mu F(x-c_t(x),t+1,T;J)\Bigg]$$

$$+ \lambda_m(t)\Bigg[m_t(x) \sum_{s=1}^{t-1} (1-\rho)^{s-1}\rho \sum_{k=1}^{t-s-1} \binom{t-s-1}{k} \rho^k$$

$$(1-\rho)^{t-s-1-k} \sum_{j=1}^{T-t-1} \binom{T-t-1}{j} \rho^j (1-\rho)^{T-t-1-j}$$

$$f(c_s^*(x)+(k+j)m^*(x)+m_t(x))$$

$$+ \mu F(x-m_t(x),t+1,T;J)\Bigg]\Bigg\}. \tag{9.34}$$

Let us first consider the terms in this equation and then understand why it is still not the correct dynamic iteration equation.

The first term on the right-hand side of Eq. (9.34) corresponds to not encountering any host in period $t$. The next term corresponds to encountering a clean host. The first term in brackets is the fitness increment from a clean host. The female lays a clutch $c_t(x)$, and in the remaining $T-t-1$ periods $j$ additional ovipositions occur with probability

$$\binom{T-t-1}{j} \rho^j (1-\rho)^{T-t-1-j}$$

resulting in the fitness increment $c_t(x)f(c_t(x)+jm^*)$ from the clutch $c_t(x)$. The second term in brackets corresponds to future fitness.

The term involving $\lambda_m(t)$ corresponds to the encounter with a marked host. The key to understanding this expression is recognizing that the probability that the first encounter of any fly with a particular head occurs in period $s$, that $k$ additional revisits occur between period $s+1$ and $t-1$, and that $j$ additional revisits occur between period $t+1$ and $T$, is given by:

$$(1-\rho)^{s-1}\rho\binom{t-1-s}{k}\rho^{k}(1-\rho)^{t-1-s-k}$$
$$\binom{T-t-1}{j}\rho^{j}(1-\rho)^{T-t-1-j}$$

and that this event leads to $c_s^* + (k+j)m^*$ eggs in the host before and after the oviposition in period $t$. This explains Eq. (9.34).

But why is Eq. (9.34) not the correct iteration equation for $F(x,t,T;J)$? Note that when ovipositions by normal (ESS) flies are considered in this equation we have written $m^*(x)$ and $c_s^*(x)$. We thus implicitly assume that in each oviposition, the normal flies have exactly the same number of eggs as the mutant does in period $t$. This surely doesn't make sense. A correct version of Eq. (9.34) must include an expectation over the distribution of egg complements of normal flies. This requires that the mutant be able to assess not only her own state, but somehow be able to estimate the state distribution of all other flies!

In order to find the distribution of eggs in normal flies, we can ignore the single mutant and—given a strategy $J$—use forward iteration to determine the probability $P_J(Z,t)$ that a fly following strategy $J$ has $Z$ eggs remaining at time $t$. We use this to take averages in Eq. (9.34) and then use the method of successive approximations to find the ESS strategy. Thus, the dynamic game is—in principle—solved, subject to the simplification (9.33). The algorithm is now

1. Input a guess for the ESS strategy $J = \{c_1^*(x), \ldots c_{T-1}^*(x), m(x)\}$.
2. Use forward iteration to find $P_J(Z,t)$.
3. Solve Eq. (9.34) with the additional expectation, as described above. If $c_t(x) = c_t^*(x)$ and $m(x) = m^*$, then stop. We have found the ESS.
4. Otherwise set $c_t^*(x) = c_t(x)$ and $m^*(x) = m(x)$ and return to step 2.

Table 9.5
Average body lengths of three behavioral morphs of juvenile coho salmon from the Salmon River in British Columbia (Puckett and Dill 1985; Table 2)

| Type | Average Length (mm) |
|---|---|
| poolers | 43 |
| territory holders | 47 |
| floaters | 37 |

This algorithm gives the solution of the dynamic game. It appears computationally feasible, with reasonable storage requirements.

### 9.2.2 A Game between Juvenile Coho Salmon*

Coho salmon (*Oncorhynchus kisutch)* spend their first year after hatching in streams. Some fish live in pools, where water velocity is relatively low, while others live in faster sections of the stream. The "poolers" are nonterritorial, whereas the fish that live in riffles consist of two types: (a) territorial fish which remain in and guard fixed areas of the stream, and (b) "floaters"—nonterritorial fish that move around in the riffles, dashing in between the occupied territories (Puckett and Dill 1985). The average body lengths of the three types observed in a particular stream are shown in Table 9.5.

A simple-minded explanation for this polymorphism in size and behavior is that larger fish are able to defend territories, while smaller fish have no alternative but to live in pools or other areas not claimed by the territorial fish. The advantage of territorial behavior is presumably a greater availability of food, which allows these fish to grow more rapidly than poolers. Large size probably confers an advantage later on, in terms of survival during migration and sea life. Larger fish may also have greater ultimate spawning success. This explanation however does not account for the fact that the smallest fish adopt the floater strategy: presumably fish of intermediate size would compete more successfully

* We thank G. Martel for suggesting this modeling problem.

with the territory holders.

To begin constructing a dynamic model, let the variable $x$ represent fish length. We treat $x$ as a continuous variable and introduce a *size distribution function* $\Psi(x,t)$, defined so that

$$\Psi(x,t)\Delta x = \text{number of fish with length between } x \text{ and } x+\Delta x \text{ in period } t$$

The total population at the start of period $t$ is

$$N(t) = \int_0^\infty \Psi(x,t)\,dx. \tag{9.35}$$

(At any given period there will be a minimum and a maximum size, $x_{\min}$ and $x_{\max}$, in the population, with $\Psi(x,t) = 0$ except for $x_{\min} \le x \le x_{\max}$. Thus $x_{\min}$ and $x_{\max}$ can be taken as the limits of integration in (9.35); however writing the limits as 0 and $\infty$ is also correct, and simplifies the notation.)

### Model 1: Growth Only

We begin with a simple static model that only considers growth and behavior of poolers and territorial fish; later we consider dynamic models that also include mortality. Suppose that for poolers, the proportional daily growth rate $G_p$ is a constant, independent of both size $x$ and the total number of poolers. The proportional growth rate $G_g(x,n_g)$ of territorial fish, however, depends both on size $x$ and on the total number of territorial fish, $n_g$ (Figure 9.4). It is reasonable to assume that

$$\frac{\partial G_g}{\partial x} > 0 \qquad \text{and} \qquad \frac{\partial G_g}{\partial n_g} \le 0 \tag{9.36}$$

i.e. larger fish benefit more from territories than small fish (because they are able to defend territories better), but the benefit decreases as the number of territorial fish increases (because of increased competition for territories). Thus the growth of territorial fish is both size and frequency dependent, while that of poolers is assumed constant.

Can we predict the total number of territorial fish? If we assume that each individual fish chooses either the territorial or the pool strategy, in such a way as to maximize its growth rate subject

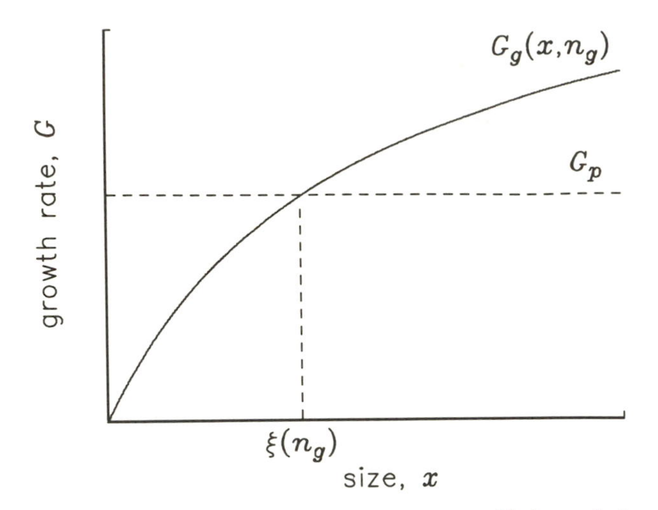


Figure 9.4 Proportional daily growth rate $G_g(x, n_g)$ for territorial fish and $G_p$ for poolers. (The significance of $\xi(n_g)$ is explained in the text.)

to how the rest of the population is distributed between pools and riffles, then the answer is yes: there is a unique equilibrium distribution of fish between pools and riffles, with the property that no individual fish can increase its growth rate by moving to the other habitat. This equilibrium is a phenotype-limited ESS, in the terminology of Parker (1984).

Let $F(x, t; I, J)$ denote the proportional growth rate of a fish of size $x$, in period $t$, if it is a single mutant using strategy $I$ in a large population of $J$-strategists. Since only growth is being considered, we equate $F(x, t; I, J)$ with the fitness increment in period $t$. Here, a strategy $J = J(x; \xi)$ has the form

$$J(x;\xi) \text{ is } \begin{cases} \text{pool} & \text{if } x \le \xi \\ \text{hold territory} & \text{if } x > \xi. \end{cases} \tag{9.37}$$

We will show that the ESS is characterized by a unique "switch point" $\bar{\xi}$, which depends on $t$. If $\bar{\xi}$ is given, then total number of territorial fish in period $t$ is

$$\bar{n}_g = \int_{\bar{\xi}}^{\infty} \Psi(x, t)\, dx. \tag{9.38}$$

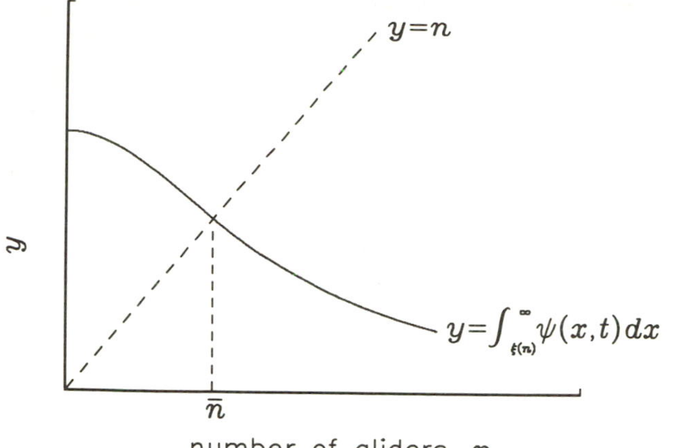


Figure 9.5 Graphical solution of Eq. (9.40): the curve $y = y(n) = \int_{\xi(n)}^{\infty} \Psi(x,t)\,dx$ is a decreasing function of $n$, because $\xi(n)$ is increasing in $n$ (see Figure 9.4). Hence there is a unique point of intersection $n = \bar{n}$ with the curve $y = n$. (We treat $n$ as a continuous variable, for simplicity.)

For any given number $n = n_g$ of territorial fish, define the size $\xi = \xi(n)$ by the equation

$$G_g(\xi, n) = G_p.$$

From Figure 9.4 we see that the fish of size $x > \xi(n)$ would do better by territorial behavior (if there are $n$ territorial fish), while those with $x < \xi(n)$ would do better by pooling.

Next, given $n_0$ territorial fish, let

$$n_1 = \int_{\xi(n_0)}^{\infty} \Psi(x,t)\,dx. \tag{9.39}$$

Thus $n_1$ equals the number of fish that do better by being territorial, when $n_0$ territorial fish are present.

When $n_1 = n_0$, no single fish can do better for itself by switching its own strategy. Setting $n_1 = n_0 = \bar{n}$ we obtain the condition

$$\bar{n} = \int_{\xi(\bar{n})}^{\infty} \Psi(x,t)\,dx \tag{9.40}$$

which characterizes the ESS switch point $\bar{\xi} = \xi(\bar{n})$, and the corresponding number of territorial fish $\bar{n}$. Equation (9.40) always has a unique solution, as explained in the caption to Figure 9.5.

Since the size distribution $\Psi(x,t)$ depends on $t$ (fish grow larger over time), the solution $\bar{n}$ to Eq. (9.40) also depends on time. Although the present model is thus a dynamic one, it is only dynamic in a very simple way. The ESS is obtained as a sequence of one-period solutions $\bar{n} = \bar{n}(t)$. This simplicity arises from our characterization of fitness solely in terms of growth—i.e. larger fish have greater ultimate fitness in terms of reproductive success.

Without specifying the exact forms of $G_p$ and $G_g(x, n_g)$, it is not possible to predict whether $\bar{n}(t)$, the number of territorial fish, would increase or decrease over time. Puckett and Dill (1985) report only average densities for the three behavioral types, but they assert that "it is quite possible that a given fish shifts between non-territorial and territorial strategies on a daily or seasonal basis" (1985, p. 98).

## Model 2: Growth and Mortality, $T = 1$

According to Puckett and Dill (1985, p.109), different feeding strategies may involve different predation risks. For example, pools tend to harbor fish predators and to provide sufficient depth for avian predators. On the other hand, the rapid movements of floaters may render them conspicuous to predators.

The modeling problem becomes much more complex when we introduce habitat and size-dependent mortality, because now a fully dynamic model is required. To get started we first consider a model with only one time period (followed by the terminal phase, when the fish migrate to sea). Such a model is appropriate if fish only make a single decision, at the beginning of the stream phase, whether to adopt territorial or pooling behavior. A more complex dynamic model is needed if fish change their decisions over time (as was the case in Model 1).

Let $\phi(x)$ denote terminal fitness for a fish of size $x$ at the end of the stream phase. Let $\sigma_p$ and $\sigma_g(x)$ denote the probabilities of survival for poolers and territorial fish, respectively; we assume that $\sigma_p$ is constant, but $\sigma_g(x)$ may depend on size $x$ (larger fish may have lower mortalities in riffles).

The lifetime fitness $F_p(x)$ of poolers (measured at the beginning of the stream phase) is given by expected terminal fitness (i.e. expected reproduction):

$$F_p(x) = \sigma_p \phi(G_p x). \tag{9.41}$$

Similarly, lifetime fitness of territorial fish, assuming that there

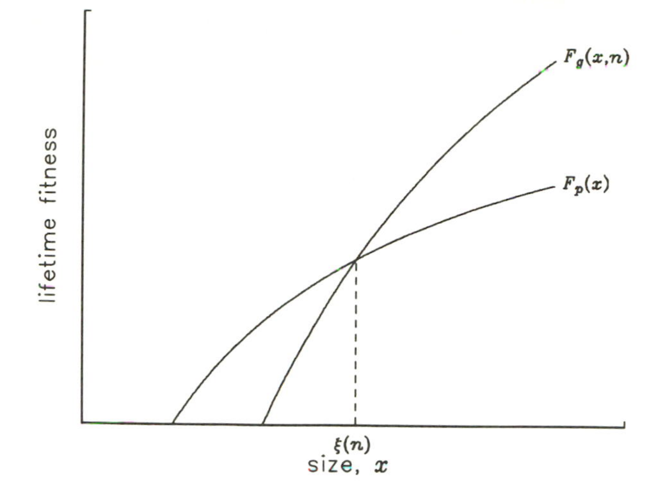


Figure 9.6 Lifetime fitness curves for territorial fish, $F_g(x, n)$, and poolers, $F_p(x)$. Fish of size greater than $\xi(n)$ do better as territory holders, and vice versa.

are $n$ territorial fish, is

$$F_g(x, n) = \sigma_g(x)\phi(G_g(x, n)x). \tag{9.42}$$

As in Model 1, the fitness of territorial fish is frequency dependent, but that of poolers is not. (This assumption is made for the sake of simplicity; it could be relaxed.)

We now consider the graphs of $F_p(x)$ and $F_g(x, n)$, and let $\xi = \xi(n)$ denote the crossover point (Figure 9.6). As in Model 1, the equilibrium solution $\bar{n}$ for the number of territorial fish is given by

$$\bar{n} = \int_{\xi(\bar{n})}^{\infty} \Psi(x)\, dx \tag{9.43}$$

where $\Psi(x)$ is the initial size distribution. Eq. (9.43) characterizes the ESS for the present one-period model; it can be solved graphically as before. (Figure 9.6 shows the case of a phenotypically mixed ESS, with some fish employing each strategy. However, the model also allows for pure ESS's—for example, one might obtain $\bar{\xi} = x_{\min}$, in which case the ESS consists entirely of territorial fish. On the other hand, if $\sigma_g(x) \ll \sigma_p$ the ESS could consist entirely of poolers.)

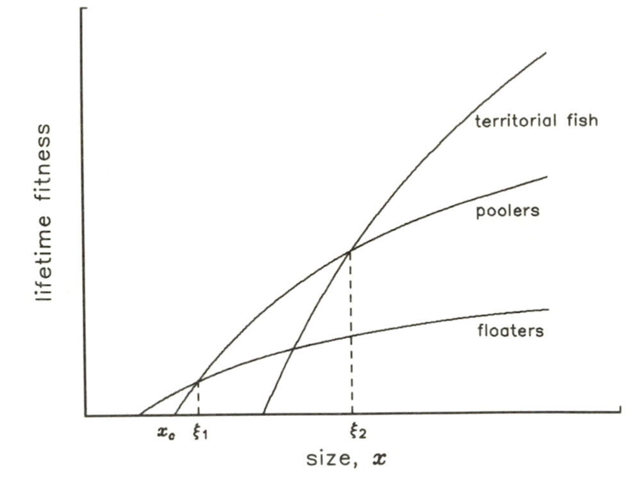


Figure 9.7 Lifetime fitness curves for territorial fish, poolers, and floaters. Fish of size greater than $\xi_2$ are territorial, those between $\xi_1$ and $\xi_2$ are poolers, and those smaller than $\xi_1$ are floaters.

The present model can be extended to include floaters. Suppose for example that $\sigma_f(x) < \sigma_p$ but $G_f(x) > G_p$: floaters grow faster than poolers, but suffer higher mortality. The fitness of floaters is given by

$$F_f(x) = \sigma_f(x)\phi(G_f(x)x). \tag{9.44}$$

This situation is depicted in Figure 9.7. The prediction of this model is that floating is the optimal strategy for small fish, pooling is optimal for intermediate sized fish, and territorial behavior is optimal for large fish (cf. Table 9.5). According to this explanation, floaters are fish which are so small, that if they adopted a pool strategy their terminal fitness would be negligible. The floating strategy, while risky, at least allows floaters a chance of achieving a viable size. That is, any fish with size below $x_c$ in Figure 9.7 has lifetime fitness identically 0 unless it adopts the floater strategy. The ESS values of $\xi_1$, $\xi_2$ can be determined by the method already outlined.

An alternative explanation for floaters is that they are opportunists who hope to displace a territorial fish, or to take over an abandoned territory, but this fails to explain why floaters are even smaller than poolers.

### Model 3: Growth and Mortality, $T > 1$

We can now set up a dynamic programming model for sequential decisions of juvenile coho salmon. Model 2 provides the solution for the penultimate period $t = T - 1$, and one might expect the generalization to $t = T - 2$, etc. to be straightforward. Unfortunately, an unavoidable technicality makes the general case vastly more difficult to solve numerically—so much so that more than three or four iterations seems beyond the capacity of the largest current supercomputer!

We begin by extending the notation for lifetime fitness:

$$F(x,t,T;\Psi(x,t);I,J) = \text{expected lifetime fitness of an } I\text{-mutant of size } x \text{ at the start of period } t, \text{ in a population of } N\ J\text{-strategists having size distribution } \Psi(x,t) \text{ at time } t. \tag{9.45}$$

Here, by strategy we mean a pair of switch points $\xi_I$, $\xi_J$, generally depending on time $t$.

We maintain that if $J$ is an ESS in the sense that for $I \neq J$

$$F(x,1,T;\Psi(x,1);I,J) < F(x,1,T;\Psi(x,1);J,J) \tag{9.46}$$

then $J$ is also an ESS from the point of view of any future time period $t$:

$$F(x,t,T;\Psi(x,t);I,J) < F(x,t,T;\Psi(x,t);J,J). \tag{9.47}$$

This is intuitively clear: if a mutant could increase its fitness by altering its strategy from period $t$ on, the same alteration would improve its lifetime fitness as of $t = 1$.

Using Eq. (9.47), we can develop an algorithm for finding the ESS. The end condition is

$$F(x,T,T;\Psi(x,T);I,J) = \phi(x). \tag{9.48}$$

For $t = T - 1$ we have the one-period problem, which can be solved as in Model 2 above. (We ignore floaters henceforth.) Let $F(x, T - 1, T; \Psi)$ denote the fitness of a fish of size $x$ using the ESS strategy in period $T - 1$:

$$F(x,T-1,T;\Psi) = \begin{cases} \sigma_P \phi(G_P x) & \text{for } x < \bar{\xi} \\ \sigma_g(x,\bar{n})\phi(G_g(x,\bar{n})x) & \text{for } x > \bar{\xi} \end{cases} \tag{9.49}$$

where $\bar{\xi}$, $\bar{n}$ denote the ESS switch point and number of territorial fish, respectively. Note that $\bar{\xi}$ and $\bar{n}$ depend on the initial distribution $\Psi = \Psi(x, T-1)$ in period $T-1$.

Next consider the case $t = T-2$. We begin with a guess $\xi_0$ for the ESS switch point in period $T-2$. Then

$$n_0 = \int_{\xi_0}^{\infty} \Psi(x, T-2)\, dx$$

represents the number of territorial fish in period $T-2$. The lifetime fitness of poolers and territorial fish are, respectively

$$\begin{aligned} F_p(x, T-2, T; \Psi) &= \sigma_p F(G_p x, T-1, T; \tilde{\Psi}) \\ F_g(x, T-2, T; \Psi) &= \sigma_g(x, n_0) F(G_g(x, n_0)x, T-1, T; \tilde{\Psi}) \end{aligned} \tag{9.50}$$

where $\tilde{\Psi} = \Psi(x, T-1)$ is the size distribution in period $T-1$ resulting from the growth and mortality that occurs in period $T-2$. The fitness functions on the right-side of Eqs. (9.50) are already determined from (9.49).

The next approximation $\xi_1$ for the switch point is obtained by solving the equation

$$F_p(x, T-2, T; \Psi) = F_g(x, T-2, T; \Psi)$$

for $x = \xi_1$. The process is repeated, leading to the next value $\xi_2$, and so on. The ESS switch point $\bar{\xi}$ is obtained as the limit of $\xi_k$ for $k \to \infty$ (but of course the computer algorithm would be coded to stop after some predetermined number of steps).

Once $\bar{\xi}$ has been determined, the fitness of the ESS is given by

$$F(x, T-2, T; \Psi) = \begin{cases} F_p(x, T-2, T; \Psi) & \text{for } x \le \bar{\xi} \\ F_g(x, T-2, T; \Psi) & \text{for } x > \bar{\xi}. \end{cases} \tag{9.51}$$

The above algorithm for $T-2$ is now iterated for $t = T-3, T-4, \ldots$, etc. Thus the ESS is obtained for all $t$.

This algorithm may seem quite straightforward, but it is in fact vastly more demanding of computer memory than any of the previous dynamic programming algorithms. For $t = T-1$ we must compute and store in memory the values of $F(x, T-1, T; \Psi)$ for all $x$ and for every possible size distribution $\Psi$. But the number of possible size distributions can be astronomical. For example, consider a small population of 100 fish, with sizes ranging from 30 to 50 mm in 1 mm steps. Then it can be shown that the total number of possible size distributions is equal to $\binom{120}{20} = 2.9 \times 10^{22}$. If a real number occupies six bytes of computer memory, the program would require over $10^{18}$ megabytes of memory. This is the curse of dimensionality with a vengeance.

There may be ways to get around this impossible memory requirement (for example, a recursive algorithm, which would compute fitness only when needed, rather than computing and storing all its values). But we will not pursue the question further. We have obtained some insight into the way frequency dependence may affect juvenile coho life history from Model 2. The purpose of discussing Model 3 was to suggest that the dynamic modeling approach advocated in this book may have some limitations, especially in the area of dynamic evolutionary games. This is probably an inescapable result of the complexity of dynamic game theory. It will be a challenge to modelers to develop simple game models that still have predictive value.

# Epilogue Perspectives on Dynamic Modeling

A reader who has been through most of this book has found that dynamic behavioral modeling is not just science. Rather, it involves the art of problem formulation, the selection of appropriate variables, constraints, and dynamics, the characterization of fitness, and finally the solution of an optimization problem. In addition, we have stressed a certain philosophical orientation towards modeling and problem-solving in biology. That is, as well as illustrating a powerful methodology, one of the goals of this book is to advocate certain perspectives for approaching biological problems. In this epilogue, we discuss some of these perspectives in more detail. We begin with a discussion of the problem of selecting the best modeling framework for a particular biological problem. We next discuss the difference between models and hypotheses. We then consider the value of constructing different models for the same phenomenon. Finally, we discuss the role of models in experimental work.

## Which Modeling Approach?

Although we have concentrated in this book on the construction and analysis of dynamic programming models of behavior, we trust that our readers will not conclude that we are recommending dynamic programming as the best, or only approach to behavioral modeling in all circumstances. As discussed in Section 8.4, many alternative modeling frameworks may be appropriate for studying a given biological phenomenon: it is always the problem rather than the technique that must be paramount in constructing a model. The modeler thus needs to be familiar with a range of techniques.

In our own modeling attempts we have often developed a sequence of different models for a particular phenomenon. Usually it turns out that one of the models eventually proves to be superior to the rest, and this is the model that gets developed and

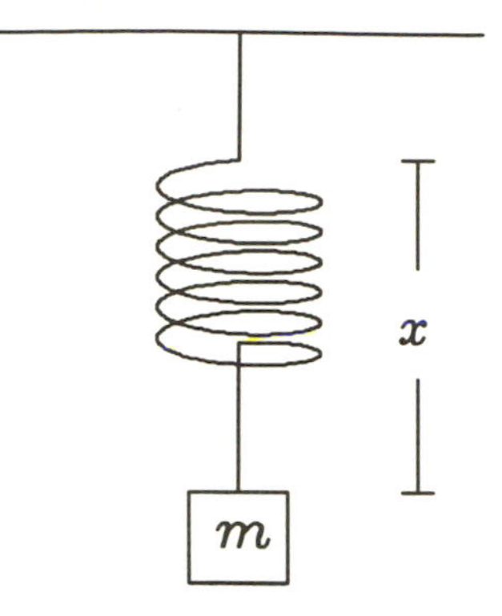


A simple model of a spring.

published. What then makes one model "better" than another? Of course this is often a matter of fine judgement; there are many conflicting desiderata, such as simplicity vs. "realism," generality vs. specificity, and so on (Fagerström 1987).

In the end, however, the touchstone of value is whether carrying out the modeling exercise has taught us anything new. A really successful model is one that suggests new explanations or hypotheses about natural phenomena, or at least indicates new ways of thinking. Such models will often suggest new experiments, which may in turn suggest new models, and so on. Just as there is seldom a final experiment, there is also seldom a final model.

## Models and Hypotheses

A model is not a scientific hypothesis. Rather, models are tools that can be used to evaluate scientific hypotheses and make predictions from these hypotheses. But the model itself is a scientific tool in the same way that laboratory equipment or field equipment is a scientific tool. It is thus important to be problem oriented, rather than tool or technique oriented.

The difference between a model and a hypothesis is nicely illustrated by a simple example from introductory physics, the motion of a spring. We let $x(t)$ denote the displacement of the spring at time $t$ (see the figure). The first spring model that people learn is

$$md^2x/dt^2 = -kx \tag{1}$$

where $k$ and $m$ are constant and $d^2x/dt^2$ denotes the second derivative of position with respect to time, i.e. acceleration. Now, Eq. (1) is a model for a physical phenomenon. It is not a scientific hypothesis, but what are the hypotheses that underlie this model? First we have Newton's second law that force equals mass times acceleration. Second, we have the hypothesis of Hooke's law, that the restoring force in a spring is proportional to the displacement. Equating these two forces (Newton's third law), we have thus constructed a simple model based on two hypotheses.

The solution of Eq. (1) is

$$x(t) = A\sin((k/m)^{1/2}t) + B\cos((k/m)^{1/2}t) \tag{2}$$

where the constants $A$ and $B$ are determined by the initial position and velocity. We start in Eq. (1) with a model based on two hypotheses, solve this model in Eq. (2), and are then able to draw insight and make predictions from the simple model. For example, we deduce by analysis of the model that the predicted motion is periodic, and that the frequency of the motion is proportional to the square root of $k$ and inversely proportional to the square root of $m$. Finally, we learn that it is not the value of $k$ or $m$ separately, but their ratio that determines the frequency of vibration. All of us have verified some of these predictions (at least qualitatively) in the past; they can be tested quantitatively in the laboratory. Actual springs do vibrate approximately according to the model—but only approximately.

Is this a "realistic model?" Certainly not. For example, we all know that springs set into motion do not oscillate forever, but stop after awhile. The model based on the original hypotheses is obviously not correct for all times—as observation times increase, observations will deviate more and more from the predictions based on this model.

What should be done about the observation that real springs eventually stop oscillating? We can modify our original model by introducing another hypothesis: frictional forces slow springs down in proportion to their velocity. Thus, we hypothesize that the frictional force is $\gamma(dx/dt)$ where $\gamma$ is the proportionality constant of the third hypothesis. Our new model, now based on three hypotheses, becomes

$$md^2x/dt^2 = -kx - \gamma(dx/dt). \quad (3)$$

The solution of this new model is

$$x(t) = A\exp(\lambda_+ t) + B\exp(\lambda_- t) \quad (4)$$

where the constants $\lambda_+$ and $\lambda_-$ are given by the formula

$$\lambda_+ = -(\gamma + (\gamma^2 - 4km)^{1/2})/2m$$

and

$$\lambda_- = -(\gamma - (\gamma^2 - 4km)^{1/2})/2m. \quad (5)$$

This model leads to new predictions. For example, if $\gamma^2 - 4km$ is less than 0, then our spring will still oscillate, but the oscillations will have exponentially decreasing amplitude, ultimately reaching 0. On the other hand, if $\gamma^2 - 4km$ is positive then our model leads to the prediction of damped motion without any oscillations at all. We also learn from the solution of this model that it is not $\gamma$ and $m$ separately, but again their ratio that determines the rate at which the spring slows down. Thus, by introducing the third hypothesis, we are able to provide more understanding about real springs.

We could continue in this vein by adding further observations about real springs, additional hypotheses, revising the model, and so forth. (As an exercise, the reader might want to try to do this. We've come up with the following additions: the force may be a function of $x$ that is nonlinear, so $-kx$ could be replaced by $F(x)$; the damping could be nonlinear, so that $-\gamma dx/dt$ could be replaced by $-\gamma_1 dx/dt - \gamma_2(dx/dt)^3$ for example; the spring could be subject to stochastic fluctuations, so that a "random force term" would be added to the right-hand side of the model; the spring could be coupled to other springs—say ten of them in a row so that the model needs to be expanded considerably. Each of these makes the mathematics much harder to do, but some new phenomena emerge out of these simple extensions: for example the theory of stochastic processes and the theory of chaos, but we will not discuss them here.) There's also an entire set of additional criticisms of this model: quantum mechanics has shown that Newtonian mechanics is "wrong," who has ever heard of a "massless spring" or a "point mass," etc.

What can we learn from this analogy in mechanics? First, throughout this book, the principle of evolution by natural selection has led us to develop models based on the concept of fitness maximization. In each situation we studied, our hypothesis concerned the components of fitness. The model was used to convert the verbal hypothesis about the components of fitness into a tool that could be used to make qualitative and quantitative predictions. We thus used the model to test the hypotheses.

Second, models help us determine what is important and what is not important for understanding the phenomenon of interest. Our models of the spring show that the main features of spring motion are essentially completely determined by three simple hypotheses and two derived parameters ($k/m$ and $\gamma/m$). By studying the model, we are able to determine key features necessary for understanding the phenomenon. For example, in Chapter 3 we studied group formation in lions. Our model showed that one key to understanding the phenomenon was the incorporation of a capacity—a single lion cannot eat an entire zebra. Perhaps this is a trivial observation, but it is so only after the fact. Models allow us to rank the importance of different factors or hypotheses about the phenomenon in a quantitative way.

## The Value of Different Models for the Same Phenomenon

Since a model is a scientific tool, there is no "correct" model. Instead, there are sequences of models, some of which may be better than others as tools for understanding certain aspects of a given phenomenon. Different models of the same phenomenon can be very useful. First (and most obviously), different models allow us to assess the validity of different hypotheses. Second, the development of more than one model often leads to a progression of understanding. Within limits, more complicated models often give better agreement with field observations, or experiments, than simple models. On the other hand, a simple model, if it provides insight, is typically much more influential in guiding thought than are more complicated models.

It is probably a natural response, upon encountering a new model, to want to reject it out of hand. Many shortcomings of the model will immediately spring to mind, especially if it is a simple model. We believe that the activity of attacking models because they are "not realistic enough" is futile. No model is ever going to be completely realistic; that is the whole point of modeling. Even the most complicated computer simulations are unable to capture all of the minutae of a particular phenomenon. The goal of modeling, as we said early on in this book, is not to reproduce nature in the computer, but to lead to understanding of natural phenomena. Instead of attacking a model with the obvious claim that "nature is more complicated than that," the creative response is to think long and hard and construct an alternative model. It is only through the construction of alternative and improved models (not merely the attack on existing ones) that we obtain a full progression of understanding.

## Predictions, Experiments, and Robustness

Enigmatic observations that cannot be understood with existing models are often the motivation for the development of new models. These new models (if they are good ones) then suggest additional variables to measure (as, for example, in the density dependence of fruit fly parasitism) and additional experiments. The cycle of experiment/model/experiment is the ideal one for modeling in biology. We believe that theoretical biology is still sufficiently young that theory should not be done in the absence of observation. But we also believe that observation without theory is often a waste of time.

When using models to make predictions for experimental measurements, two important issues arise. The first is that in the vicinity of the optimum of the fitness function, the fitness function is relatively flat. This is seen from elementary calculus. Suppose that $f(x)$ is a function with a maximum near $x_m$. Then in the vicinity of $x_m$, we have by Taylor's expansion

$$\begin{aligned} f(x) &\cong f(x_m) + f'(x_m)(x - x_m) + (1/2)f''(x_m)(x - x_m)^2 \\ &= f(x_m) + (1/2)f''(x_m)(x - x_m)^2 \end{aligned}$$

since $f'(x_m) = 0$ by the assumption that $x_m$ gives a maximum of the function $f(x)$. Thus, small deviations in $x$ from the maximizing value $x_m$ lead to even smaller changes in $f(x)$. We saw this property with our prediction of clutch sizes in Section 6.1. For the model developed there, clutches of five were optimal in terms of expected number of offspring surviving, but clutches of three and four were so close to optimal that our prediction was that clutches in the range of 3–5 would be observed.

The second issue concerns the matter of sensitivity of the model to changes in assumptions and parameters. A robust model should have the same qualitative properties over a wide range of parameter values. It is often worthwhile to try to understand what features of a particular model are responsible for its main properties. For example, the simple spring model of Eq. (1) predicts that vibrations will continue unabated indefinitely. The discussion following Eq. (3) shows that it is the assumption of zero friction that leads to this unrealistic prediction. With the new model we are able to quantify the effects of friction. Thus the prediction of vibration undiminished over time is not a robust prediction of the original model.

Let us consider another example, the model of vertical migration of planktivores discussed in Section 5.2. First, we constructed a very simplistic submodel of aquatic predation to deduce the existence of an "antipredation window" at dawn and dusk. Using this submodel, we were able to construct a dynamic model of vertical migration that successfully predicted the observations. However, the predation submodel is obviously extremely unrealistic; the effects of changes in light intensity on fish vision are known to be highly complex. How robust is the vertical migration model, given that it is based on an unrealistic model of aquatic predation?

The critical hypothesis underlying the vertical migration model is the existence of brief intervals at dawn and dusk during which predation risks are low relative to potential feeding rates. As noted in Chapter 5, this hypothesis could be tested independently in the lab. (Once again we see the progression of observations, modeling, and experiment, which may be followed by model revision, further experiments, and so on.) If the experiments turn out to support the hypothesis, then the explanation of vertical migration as a response to differential predation risks and feeding opportunities would be strengthened, regardless of the oversimplicity of the original model with respect to the effects of illumination on predation rates.

Any component of a model that affects qualitative behavior in a major way is a key component, and its effects could be referred to as first-order effects. Components that only cause minor quantitative effects may be referred to as having a second-order influence on the model's predictions. One of the advantages of the computer-oriented approach to behavioral modeling advocated in this book is that computer "experiments" (often called sensitivity tests) can readily be made, and the first- and second-order effects identified.

For example, in the vertical migration model, the existence of an antipredation window is a first-order component, but the precise visual mechanism underlying this component would be of second order.

Similarly in the model of lions in Chapter 3, the inclusion of gut capacity is a first-order effect, but allowing differential metabolic costs for foraging or protecting a carcass would be a second-order effect. When using models to make predictions, we want to concentrate mainly on first-order effects since they will most easily allow us to test the validity of particular hypotheses.

## Computers, Biology, and Modeling

There is a saying that can usually be found in at least one office in the Computer Science Department of any university: "The purpose of computing is insight, not numbers." Most of the models that we have developed in this book require computer solution. It is true that essentially all of them can be solved on small desktop microcomputers, but even so there is the apparent limitation of the computer-oriented approach that it is not possible to obtain sharp, neat analytic results. It may appear that it is impossible to derive general principles in behavioral ecology using the computer. We have three comments.

First, dynamic optimization models are notoriously difficult to solve analytically. There are some special cases (e.g. linear dynamics and quadratic costs) in which general solutions are known. But these special cases are not robust to small changes in the model (e.g. a small nonlinear term in either the dynamics or cost function invalidates the entire analytical solution). In writing this book, we were unwilling, and hope that readers agree, to force behavioral modeling into a particular methodological straightjacket. Models that have been highly successful in physics or engineering need not be the best models in some other field, especially biology. A model may have a "life of its own" to a mathematician, but its value in science is determined solely by its ability to improve our understanding and prediction of natural phenomena.

Second, we believe that it is possible to deduce general principles from computer examples, e.g. the role of gut capacity in the formation of groups of lions, the role of egg limitation in the oviposition decisions of insects, or the role of the antipredation window in aquatic biology. The very process of constructing a good dynamic optimization model often leads to new insights about the phenomenon (e.g. the movement of the spiders at the creek and lake habitats), and can possibly lead to general principles. In fact, one of the most useful roles of theory in biology is to help organize one's thinking about a particular problem. In many cases (e.g. lions, planktivores, spiders) numerical computation is just "icing on the cake," once the model has been developed.

Third, the use of a computer is a double-edged sword. It is true that neat analytical results are not possible. On the other hand, once the model is constructed and realized as a computer program, we are able to conduct computer-based "experiments" by either modifying parameters or using Monte Carlo simulations (e.g. the response of insects to host deprivation, or the population consequences of natal dispersal). Such experiments may lead to the formation of general principles, although a pure theoretician might complain that they lack logical rigor. Finally, such computer experiments can suggest laboratory experiments or field observations.

To summarize our viewpoint, the world of biology is complex, dynamic, and stochastic. The state variable, dynamic programming approach described in this book is an attempt to come to grips with this reality, in a framework that remains relatively transparent from both biological and mathematical viewpoints. Using the computer to determine optimal (or evolutionarily stable) behavioral sequences for our models frees us from the need for highly technical mathematical analyses. As with any modeling technique, the limitations of this approach must be recognized and admitted. But, when used with discretion, we believe that dynamic state variable models have the potential for permanently affecting the ways that evolutionary biologists think about the interplay between behavior and ecology.

# References


Abramowitz, M. and I.A. Stegun. 1965. *Handbook of Mathematical Functions.* National Bureau of Standards, U.S. Dept. of Commerce, Washington, D.C.

Allee, B.A. 1981. The role of interspecific competition in the distribution of salmonids in streams. In E.L. Brannon and E.O. Salo (eds.), *Salmon and Trout Migratory Behavior Symposium.* School of Fisheries, Univ. of Wash., Seattle, WA, pp. 111–122.

Aoki, M. 1967. *Optimization of Stochastic Systems.* Academic Press, New York.

Armitage, K.B. 1986. Individuality, social behavior, and reproductive success in yellow-bellied marmots. *Ecology* 67: 1186–1193.

Bachman, R.A. 1981. A growth model for drift-feeding salmonids: A selective pressure for migration. In E.L. Brannon and E.O. Salo (eds.), *Salmon and Trout Migratory Behavior Symposium.* School of Fisheries, Univ. of Wash., Seattle, WA, pp. 128–135.

Baker, R.R. 1978. *The Evolutionary Ecology of Animal Migration.* Holmes and Meier, New York.

Basar, T. and G.J. Olsder. 1982. *Dynamic Noncooperative Game Theory.* Academic Press, New York.

Bellman, R. 1957. *Dynamic Programming.* Princeton Univ. Press, Princeton, NJ.

Berger, J.O. 1980. *Statistical Decision Theory.* Springer–Verlag, Heidelberg.

Bertram, B.C.R. 1975. Social factors influencing reproduction in wild lions. *J. Zool., Lond.* 177: 463–482.

Bertram, B.C.R. 1982. Leopard ecology as studied by radio tracking. *Symp. Zool. Soc. Lond.* 49: 341–352.

Blaxter, J.H.S. 1974. The role of light in the vertical migration of fish—a review. In G.C. Evans, R. Bainbridge and O. Rackham (eds.), *Light as an Ecological Factor: II.* British Ecological Society 16th Symposium. Blackwell, Oxford, pp. 189–210.

Borowicz, V.A. and S.A. Juliano. 1986. Inverse density dependent parasitism of Cornus amomum fruitfly *Rhagoletis cornivora. Ecology* 67: 639–643.

Boyce, M.S. and C.M. Perrins. 1987. Optimizing great tit clutch size in a fluctuating environment. *Ecology* 68: 142–153.

Brian, M.V. 1983. *Social Insects.* Chapman and Hall, N.Y.

Brown, J.L. 1969. Territorial behavior and population regulation in birds. *Wilson Bull.* 81: 293–329.

Brown, J.L. 1987. *Helping and Communal Breeding in Birds.* Princeton Univ. Press, Princeton, NJ.

Bryant, D.M. and A.K. Turner. 1982. Central place foraging by swallows (hirundinidae): the question of load size. *Anim. Behav.* 30: 845–856.

Burtt, E.H. 1977. Some factors in the timing of parent–chick recognition in swallows. *Anim. Behav.* 25: 231–239.

Caldwell, G. 1986. Predation as a selective force on foraging herons: effects of plumage color and flocking. *Auk* 103: 494–505.

Caraco, T. 1981. Risk sensitivity and foraging groups. *Ecology* 62: 527–531.

Caraco, T. and R.G. Gillespie. 1986. Risk-sensitivity: foraging mode in an ambush predator. *Ecology* 67: 1180–1185.

Caraco, T. and L.L. Wolf. 1975. Ecological determinants of group sizes of foraging lions. *Amer. Nat.* 109: 343–352.

Caraco, T., S. Martindale, and T.S. Whittam. 1980. An empirical demonstration of risk-sensitive foraging preferences. *Anim. Behav.* 28: 820–830.

Carey, J.R. 1984. Host-specific demographic studies of the Mediterranean fruit fly *Ceratitis capitata. Ecol. Entomol.* 9: 261–270.

Carlson, A. and J. Moreno. 1981 . Central place foraging in the wheatear *Oenanthe oenanthe*: an experimental test. *J. Anim. Ecol.* 50: 917–924.

Cerri, R.D. 1983. The effect of light intensity on predator and prey behaviour in cyprinid fish: factors that influence prey risk. *Anim. Behav.* 31: 736–742.

Cerri, R.D. and D.F. Fraser. 1983. Predation and risk in foraging minnows: balancing conflicting demands. *Amer. Nat.* 121: 552–561.

Charnov, E.L. 1976. Optimal foraging, the marginal value theorem. *Theor. Pop. Biol.* 9: 129–136.

Charnov, E.L. 1982. *The Theory of Sex Allocation.* Princeton Univ. Press, Princeton, NJ.

Charnov, E.L. and S.W. Skinner. 1984. Evolution of host selection and clutch size in parasitoid wasps. *Florida Entomol.* 67: 5–21.

Charnov, E.L. and S.W. Skinner. 1985. Complementary approaches to understanding parasitoid oviposition decisions. *Environ. Entomol.* 14: 383–391.

Clark, C.W. 1985. *Bioeconomic Modelling and Fisheries Management.* Wiley–Interscience, New York.

Clark, C.W. 1987. The lazy, adaptable lions: a Markovian model of group foraging. *Anim. Behav.* 35: 361–368.

Clark, C.W. and R.H. Lamberson. 1988. Polymorphic reproductive strategies in stochastic environments. *Theor. Pop. Biol.* (in press).

Clark, C.W. and D.A. Levy. 1988. Diel vertical migrations by juvenile sockeye salmon. *Amer. Nat.* 131: 271–290.

Clark, C.W. and M. Mangel. 1984. Foraging and flocking strategies: information in an uncertain environment. *Amer. Nat.* 123: 626–641.

Clark, C.W. and M. Mangel. 1986. The evolutionary advantages of group foraging. *Theor. Pop. Biol.* 30: 45–75.

Cohen, D. 1966. Optimizing reproduction in a randomly varying environment. *J. Theor. Biol.* 12: 119–129.

Courtney, S.P. 1986. Why insects move between host patches: some comments on 'risk-spreading.' *Oikos* 47: 112–114.

Cunjak, R.A. and G. Power. 1986. Winter habitat utilization by stream resident brook trout (*Salvelinus fontinalis*) and brown trout (*Salmo trutta*). *Can. J. Fish. Aquat. Sci.* 43: 1970–1981.

David, P.M. 1961. The influence of vertical migration on speciation in the oceanic plankton. *Syst. Zool.* 10: 10–16.

Dawkins, R. 1976. *The Selfish Gene.* Oxford Univ. Press, Oxford.

De Coursey, P.J. (ed.). 1976. *Biological Rhythms in the Marine Environment.* Univ. South Carolina Press, Columbia, SC.

de Groot, M. 1970. *Optimal Statistical Decisions.* McGraw–Hill, New York.

De Jong, G. 1979. The influence of the distribution of juveniles over patches of food on the dynamics of a population. *Nether. J. Zool.* 29: 33–51.

De Jong, G. 1982. The influence of dispersal pattern on the evolution of fecundity. *Nether. J. Zool.* 32: 1–30.

Derr, J.A., B. Alden, and H. Dingle. 1981. Insect life histories in relation to migration, body size, and host plant array: a comparative study of *Dysdercus*. *J. Anim. Ecol.* 50: 181–193.

Dethier, V.G. 1982. Mechanism of host-plant recognition. *Ent. Expl. & Appl.* 31: 49–56.

Dingle, H. 1984. Behavior, genes, and life histories: complex adaptations in uncertain environments. In P.W. Price, C.N. Slobodchikoff, and W.S. Gaud (eds.), *A New Ecology.* Wiley-Interscience, New York, pp. 169–194.

Drent, R.H. and S. Daan. 1980. The prudent parent: energetic adjustments in avian breeding. *Ardea* 68: 225–252.

Dubins, L.E. and L.J. Savage. 1976. *Inequalities for Stochastic Processes (How to Gamble If You Must).* Dover Publications, New York.

Dumont, H.J. 1972. A competition-based approach to the reverse vertical migration in zooplankton and its implications, chiefly based on a study of the interactions of the rotifer *Asplanchna priodonta* (Gosse) with several crustacea entomostraca. *Int. Rev. Ges. Hydrobiol.* 57: 1–38.

Eggers, D.M. 1976. Theoretical effects of schooling by planktivorous fish predators on the rate of prey consumption. *J. Fish. Res. Board Can.* 33: 1964–1971.

Ekman, J. and C. Askenmo. 1986. Reproductive cost, age-specific survival and a comparison of the reproductive strategy in two European tits (Genus *Parus*). *Evolution* 40: 159–168.

Ellner, S. 1985. ESS germination strategies in randomly varying environments. I. Logistic-type models. *Theor. Pop. Biol.* 28: 50–79.

Emlen, J.M. 1966. The role of time and energy in food preference. *Amer. Nat.* 100: 611–617.

Enright, J.T. 1977. Diurnal vertical migration: Adaptive significance and timing. Part I. Selective advantage: A metabolic model. *Limn. Oceanogr.* 22: 856–872.

Fagerström, T. 1987. On theory, data and mathematics in ecology. *Oikos* 50: 258–261.

Fisher, R.A. 1930. *The Genetical Theory of Natural Selection.* Clarendon Press, Oxford.

Fitt, G.P. 1984. Oviposition behaviour of two tephritid fruit flies, *Dacus tryoni* and *Dacus jarvisi*, as influenced by the presence of larvae in the host fruit. *Oecologia* 62: 37–46.

Foerster, R.E. 1968. The sockeye salmon, *Oncorhynchus nerka*. *Bull. Fish. Res. Board Can.* 162: 422p.

Fraser, D.F. and R.D. Cerri. 1982. Experimental evaluation of predator–prey relationships in a patchy environment: consequences for habitat use patterns in minnows. *Ecology* 63: 307–313.

Gillespie, R.G. and T. Caraco. 1987. Risk-sensitive foraging strategies of two spider populations. *Ecology* 68: 887–899.

Gilliam, J.F. 1982. Habitat use and competitive bottlenecks in size-structured fish populations. Ph. D. diss., Michigan State University.

Giraldeau, L.-A. 1986. The stable group and the determinants of foraging group size (unpublished manuscript).

Gliwicz, Z.M. 1986a. Predation and the evolution of vertical migration in zooplankton. *Nature* 320: 746–748.

Gliwicz, Z.M. 1986b. A lunar cycle in zooplankton. *Ecology* 67: 883–897.

Goodman, D. 1982. Optimal life histories, optimal notation, and the value of reproductive value. *Amer. Nat.* 119: 803–823.

Gossard, T.W. and R.E. Jones. 1977. The effects of age and weather on egg laying in *Pieris rapae* L. *J. Appl. Ecol.* 14: 65–71.

Gould, S.J. and R.C. Lewontin. 1979. The spandrels of San Marco and the Panglossian paradigm: a critique of the adaptationist programme. *Proc. Roy. Soc. London. B.* 205: 581–598.

Grafen, A. 1984. Natural selection, kin selection, and group selection. In J.R. Krebs and N.B. Davies (eds.), *Behavioural Ecology.* Second edition. Blackwell, Oxford, pp. 62–84.

Greenwood, P.J., P.H. Harvey, and C.M. Perrins. 1979. The role of dispersal in the great tit (*Parus major*): the causes, consequences and heritability of natal dispersal. *J. Anim. Ecol.* 48: 123–142.

Grossmueller, D.W. and R.C. Lederhouse. 1985. Oviposition site selection: an aid to rapid growth and development in the tiger swallowtail butterfly, *Papilio galucus. Oecologia* 66: 68–73.

Grubb, T.C. and L. Greenwald. 1982. Sparrows and a brushpile: foraging responses to different combinations of predation risk and energy cost. *Anim. Behav.* 30: 637–640.

Hairston, N.G. 1976. Photoprotection by carotenoid pigments in the copepod *Diaptomus nevadensis. Proc. Nat. Acad. Sci. USA* 73: 971–974.

Hairston, N.G. and W.R. Munns. 1984. The timing of copepod diapause as an evolutionarily stable strategy. *Amer. Nat.* 123: 733–751.

Hall, D.J., E.E. Werner, J.F. Gilliam, G.G. Mittelbach, D. Howard, C.G. Doner, J.A. Dickerman, and A.J. Stewart. 1979. Diel foraging behavior and prey selection in the golden shiner (*Notemigonus crysoleucas*). *J. Fish. Res. Board Can.* 36: 1029–1039.

Hamilton, W.D. 1964. The genetical evolution of social behaviour. *J. Theor. Biol.* 7: 1–52.

Hamilton, W.D. 1967. Extraordinary sex ratios. *Science* 156: 477–488.

Hanby, J.P. and J.D. Bygott. 1987. Emigration of subadult lions. *Anim. Behav.* 35: 161–169.

Hardy, J.E. 1953. Physical factors involved in the vertical migration of plankton. *Quart. J. Microscop. Sci.* 94: 537–550.

Hardy, A.C. 1958. *The Open Sea: Its Natural History Part 1: The World of Plankton.* Collins, London.

Hardy, A.C. and E.R. Gunther. 1935. The plankton of the South Georgia whaling grounds and adjacent waters, 1926-1927. *Discovery Report* 11: 1–456.

Hassell, M.P. 1986. Parasitoids and population regulation. In J. Waage and D. Greathead (eds.), *Insect Parasitoids.* Academic Press, New York, pp. 201–224.

Hayes, J.L. 1985. Egg distribution and survivorship in the pierid butterfly *Colias alexandra. Oecologia* 66: 495–498.

Hayes, J.W. 1987. Competition for spawning space between brown (*Salmo trutta*) and rainbow trout (*S. gairdneri*) in a lake inlet tributary, New Zealand. *Can. J. Fish. Aquat. Sci.* 44: 40–47.

Hegner, R.E. 1982. Central place foraging in the white-fronted bee-eater. *Anim. Behav.* 30: 953–963.

Heiner, R. 1983. The origin of predictable behavior. *Amer. Econ. Rev.* 73: 560–595.

Henderson, M.A. and T.G. Northcote. 1985. Visual prey detection and foraging in sympatric cutthroat trout (*Salmo clarki clarki*) and Dolly Varden (*Salvelinus malma*). *Can. J. Fish. Fish. Aquat. Sci.* 42: 785–790.

Heyman, D.P. and M.J. Sobel. 1984. *Stochastic Models in Operations*

*Research*. Vol. 2. McGraw Hill, New York.

Holekamp, K.E. 1986. Proximal causes of natal dispersal in Belding's ground squirrels (*Spermophilus beldingi*). *Ecol. Monogr.* 56: 365–391.

Holling, C.S. 1965. The functional response of predators to prey density and its role in mimicry and population regulation. *Mem. Ent. Soc. Can.* 45: 1–60.

Houston, A.I. and J.M. McNamara. 1985. The choice of two prey types that minimises the probability of starvation. *Behav. Ecol. Sociobiol.* 17: 135–141.

Houston, A.I. and J.M. McNamara. 1986. Evaluating the selection pressure on foraging decisions. In R. Chapman and R. Zyan (eds.), *Relevance of Models and Theories in Ethology*. Privat, I.C.C., Toulouse, pp. 61–75.

Houston, A.I. and J.M. McNamara. 1987. Singing to attract a mate—a stochastic dynamic game. *J. Theor. Biol.* 129: 57–68.

Huey, R.B. and E.R. Pianka. 1981. Ecological consequences of foraging mode. *Ecology* 62: 991–999.

Iwasa, Y. 1982. Vertical migration of zooplankton: a game between predator and prey. *Amer. Nat.* 120: 171–180.

Iwasa, Y., Y. Suzuki, and H. Matsuda. 1984. Theory of oviposition strategy of parasitoids. I. Effect of mortality and limited egg number. *Theor. Pop. Biol.* 26: 205–227.

Jenkins, T.M. 1969. Social structure, position choice, and micro-distribution of two trout species (*Salmo trutta* and *Salmo gairdneri*) resident in mountain streams. *Animal Behavior Monographs* 2: 57–123.

Johnsgard, P.A. 1973. Proximate and ultimate determinants of clutch size in Anatidae. *Wildfowl* 24: 144–149.

Johnston, J.M. 1981. Life histories of anadromous cutthroat with emphasis on migratory behavior. In E.L. Brannon and E.O. Salo (eds.), *Salmon and Trout Migratory Behavior Symposium*. School of Fisheries, Univ. of Wash., Seattle, WA, pp. 123–127.

Jones, R.E. 1977. Movement patterns and egg distribution in cabbage butterflies. *J. Anim. Ecol.* 46: 195–212.

Kacelnik, A. 1984. Central place foraging in starlings (*Sturnus vulgaris*). I. Patch residence time. *J. Anim. Ecol.* 53: 283–299.

Kacelnik, A. 1986. Short-term adjustments of parental effort in starlings. In H. Ouellet (ed.), *Proc. 21st Ornith. Conf.*, Ottawa, Canada.

Kahneman, D. and A. Tversky. 1979. Prospect theory: an analysis of decision under risk. *Econometrica* 47: 263–291.

Kamien, M.I. and N.L. Schwartz. 1981. *Dynamic Optimization: The Calculus of Variations and Optimal Control in Economics and Management*. North Holland, New York.

Karlin, S. and S. Lessard. 1986. *Theoretical Studies on Sex Ratio*

*Evolution*. Princeton Univ. Press, Princeton, NJ.

Karlin, S. and H.M. Taylor. 1977. *A First Course in Stochastic Processes*. Academic Press, New York.

Katz, P.L. A long-term approach to foraging optimization. *Amer. Nat.* 108: 758–782.

Kerfoot, W.B. 1970. Bioenergetics of vertical migration. *Amer. Nat.* 104: 529–546.

Kiester, A.R. and M. Slatkin. 1974. A strategy of movement and resource utilization. *Theor. Pop. Biol.* 6: 1–20.

Klomp, H. 1970. The determination of clutch-size in birds. A review. *Ardea* 58: 1–124.

Klomp, H. and B.J. Teernik. 1967. The significance of oviposition rates in the egg parasite *Trichogramma embryophagum*. *Arch. Neerl. Zool.* 17: 350–373.

Krebs, J.R. and N.B. Davies. 1984. *Behavioural Ecology*. Second edition. Blackwell, Oxford.

Krebs, J.R. and R.H. McCleery. 1984. Optimization in behavioral ecology. In J.R. Krebs and N.B. Davies (eds.), *Behavioural Ecology*. Second edition. Blackwell, Oxford, pp. 91–121.

Krebs, J.R., D.W. Stephens, and J. Sutherland. 1983. Perspectives in optimal foraging. In G.A. Clark and A.H. Brush (eds.), *Perspectives in Ornithology*. Cambridge Univ. Press, Cambridge, New York, pp. 165–221.

Kurland, J.A. and S.J. Beckerman. 1985. Optimal foraging and hominid evolution: labor and reciprocity. *Amer. Anthrop.* 87: 73–93.

Lack, D. 1954. *The Natural Regulation of Animal Numbers*. Clarendon Press, Oxford.

Lack, D. 1966. *Population Studies of Birds*. Oxford Univ. Press, Oxford.

Lack. D. 1968. *Ecological Adaptations for Breeding in Birds*. Methuen, London.

Lane, P.A. 1975. The dynamics of aquatic systems: a comparative study of the structure of four zooplankton communities. *Ecol. Monogr.* 45: 307–336.

Le Boeuf, B.J. and B.R. Mate. 1978. Elephant seals colonize additional Mexican and Californian islands. *J. Mamm.* 59: 621–622.

Le Boeuf, B.J. and K.J. Panken. 1977. Elephant seals breeding on the mainland in California. *Proc. Calif. Acad. Sci., Fourth Series* 41: 267–280.

Leon, J.A. 1976. Life histories as adaptive strategies. *J. Theor. Biol.* 60: 301–335.

Levin S.A., A. Hastings, and D. Cohen. 1984. Dispersal strategies in patchy environments. *Theor. Pop. Biol.* 26: 165–191.

Levy, D.A. 1987. Review of the ecological significance of diel vertical migrations by juvenile sockeye salmon *Oncorhynchus nerka*. *Can.*

*Spec. Publ. Fish. Aquat. Sci.* 96: 44–52.

Lewontin, R.C. and D. Cohen. 1969. On population growth in a randomly varying environment. *Proc. Nat. Acad. Sci. USA* 62: 1056–1060.

Lima, S.L. 1984. Downy woodpecker foraging behavior: efficient sampling in simple stochastic environments. *Ecology* 65: 166–174.

Lima, S.L. 1985a. Maximizing feeding efficiency and minimizing time exposed to predators: a trade-off in the black-capped chickadee. *Oecologia* 66: 60–67.

Lima, S.L. 1985b. Sampling behavior of starlings foraging in simple patchy environments. *Behav. Ecol. Sociobiol.* 16: 135–142.

Lomnicki, A. 1978. Individual differences between animals and the natural regulation of their numbers. *J. Anim. Ecol.* 47: 461–475.

Lomnicki, A. 1988. *Population Ecology of Individuals.* Princeton Univ. Press, Princeton, NJ.

Lynch, M. 1980. The evolution of cladoceran life histories. *Quart. Rev. Biology* 55: 23–42.

MacArthur, R.H. and E.R. Pianka. 1966. On the optimal use of a patchy environment. *Amer. Nat.* 100: 603–609.

Machina, M.J. 1987. Decision-making in the presence of risk. *Science* 236: 537–543.

Mackauer, M. 1982. Fecundity and host utilization of the aphid parasite *Aphelinus semiflavus* (Hymenoptera: Aphelinidae) at two host densities. *Can. Ent.* 114: 721–726.

Mackintosh, N.A. 1937. The seasonal circulation of the Antarctic macroplankton. *Discovery Report* 16: 365–412.

Mangel, M. 1985a. Search models in fisheries and agriculture. In M. Mangel (ed.), Springer–Verlag *Lect. Notes in Biomath.* 61: 105–138.

Mangel, M. 1985b. *Decision and Control in Uncertain Resource Systems.* Academic Press, New York.

Mangel, M. 1987a. Modelling behavioral decisions of insects. In Y. Cohen (ed.), Springer-Verlag *Lect. Notes in Biomath.* 73: 1–18.

Mangel, M. 1987b. Oviposition site selection and clutch size in insects. *J. Math. Biol.* 25: 1–22.

Mangel, M. and C.W. Clark. 1983. Uncertainty, search, and information in fisheries. *J. Cons. int. Explor. Mer* 41: 93–103.

Mangel, M. and C.W. Clark. 1986. Towards a unified foraging theory. *Ecology* 67: 1127–1138.

Marquiss, M. and I. Newton. 1981. A radio-tracking study of the ranging behaviour and dispersion of European sparrowhawks *Accipiter nisus. J. Anim. Ecol.* 51: 111–133.

Marris, G., S. Hubbard, and J. Hughes. 1986. Use of patchy resources by *Nemeritis canescens* (Hymenoptera: Ichneumonidae). I. Optimal solutions. *J. Anim. Ecol.* 55: 631–640.

Martz, H.F. and R.A. Waller. 1982. *Bayesian Reliability Analysis*. Wiley, New York.

Maynard Smith, J. 1982. *Evolution and The Theory of Games*. Cambridge Univ. Press, Cambridge.

McDonald, P.T. and D.O. McInnis. 1985. *Ceratitis capitata*: Effect of host fruit size on the number of eggs per clutch. *Ent. Expl. & Appl.* 37: 207–211.

McFarland, D. (ed.). 1982. *Functional Ontogeny*. Pitman, London.

McFarland, D. and A.I. Houston. 1981. *Quantitative Ethology: The State Space Approach*. Pitman, London.

McFarland, W.N., J.C. Ogden, and J.N. Lythgoe. 1979. The influence of light on the twilight migrations of grunts. *Env. Biol. Fish.* 4: 9–22.

McLaren, I.A. 1963. Effects of temperature on growth of zooplankton, and the adaptive value of vertical migration. *J. Fish. Res. Bd. Canada* 20: 685–727.

McLaren, I.A. 1974. Demographic strategy of vertical migration by a marine copepod. *Amer. Nat.* 108: 91–102.

McNamara, J.M. and A.I. Houston. 1982. Short term behavior and lifetime fitness. In D.J. McFarland (ed.), *Functional Ontogeny*. Pitman, London, pp. 60–87.

McNamara, J.M. and A.I. Houston. 1986. The common currency for behavioural decisions. *Amer. Nat.* 127: 358–378.

McNamara, J.M., R.H. Mace, and A.I. Houston. 1987. Optimal daily routines of singing and foraging in a bird singing to attract a mate. *Behav. Ecol. Sociobiol.* 20: 399–405.

Mesterton-Gibbons, M. 1988. On the optimal compromise for a dispersing parasitoid. *J. Math. Biol.* (in press).

Milinski, M. 1986. Constraints placed by predators on feeding behaviour. In T.J. Pitcher (ed.), *The Behaviour of Teleost Fishes*. Croom Helm, London, pp. 236–252.

Milinski, M. and R. Heller. 1978. Influence of a predator on the optimal foraging behaviour of sticklebacks (*Gasterosteus aculeatus* L.). *Nature* 275: 642–644.

Miller, R.J. and E.L. Brannon. 1981. The origin and development of life history patterns in Pacific salmonids. In E.L. Brannon and E.O. Salo (eds.), *Salmon and Trout Migratory Behavior Symposium*. School of Fisheries, Univ. of Wash., Seattle, WA, pp. 296–309.

Mittelbach, G.G. 1981. Foraging efficiency and body size: a study of optimal diet and habitat use by bluegills. *Ecology* 62: 1370–1386.

Moore, H.B. and E.G. Corwin. 1956. The effects of temperature, illumination and pressure on the vertical distribution of zooplankton. *Bull. Mar. Sci. Gulf Caribbean* 6: 273–287.

Morrison, G. and W.J. Lewis. 1981. The allocation of searching time by *Trichogramma pretiosum* in host-containing patches. *Ent. Expl.*

*& Appl.* 30: 31–39.

Murphy, G.I. 1968. Pattern in life history and the environment. *Amer. Nat.* 102: 391–403.

Narver, D.W. 1970. Diel vertical movements and feeding of underyearling sockeye salmon and the limnetic zooplankton in Babine Lake, British Columbia. *J. Fish. Res. Board Can.* 27: 281–316.

Newton, I. 1986. *The Sparrowhawk.* T & A D Poyser, Town Head House, Calton, Waterhouses, Staffordshire, England.

Newton, I. and M. Marquiss. 1983. Dispersal of sparrowhawks between birthplace and breeding place. *J. Anim. Ecol.* 52: 463–477.

Nur, N. 1984a. The consequences of brood size for breeding blue tits. I. Adult survival, weight change and the cost of reproduction. *J. Anim. Ecol.* 53: 479–496.

Nur, N. 1984b. The consequences of brood size for breeding blue tits. II. Nestling weight, offspring survival and optimal brood size. *J. Anim. Ecol.* 53: 497–517.

Nur, N. 1986. Is clutch size variation in the blue tit (*Parus caeruleus*) adaptive? An experimental study. *J. Anim. Ecol.* 55: 983–999.

Oaten, A. 1977. Optimal foraging in patches: a case for stochasticity. *Theor. Pop. Biol.* 12: 263–285.

O'Connor, R.J. 1978. Brood reduction in birds: selection for fratricide, infanticide, and suicide? *Anim. Behav.* 26: 79–96.

Ohman, M.D., B.W. Frost, and E.B. Cohen. 1983. Reverse diel vertical migration: an escape from invertebrate predators. *Science* 220: 1404–1407.

Oster, G.F. and E.O. Wilson. 1978. *Caste and Ecology in the Social Insects.* Princeton Univ. Press, Princeton, NJ.

Owens, M. and D. Owens. 1984. *Cry of the Kalahari.* Houghton Mifflin, Boston, MA.

Packer, C. 1986. The ecology of sociality in felids. In D.I. Rubenstein and R.W. Wrangham (eds.), *Ecological Aspects of Social Evolution.* Princeton Univ. Press, Princeton, NJ, pp. 429–451.

Packer, C., L. Herbst, A.E. Pusey, J.D. Bygott, J.P. Hanby, S.J. Cairns and M. Borgerhoff-Mulder. 1987. Reproductive success of lions. In T.H. Clutton-Brock (ed.), *Reproductive Success*, Univ. of Chicago Press, Chicago, IL.

Parker, G.A. 1984. Evolutionary stable strategies. In J.R. Krebs and N.B. Davies (eds.), *Behavioural Ecology.* Second edition. Blackwell, Oxford, pp. 30–61.

Patterson, I.J. 1980. Territorial behaviour and the limitation of population density. *Ardea* 68: 53–62.

Pielou, E.C. 1977. *Mathematical Ecology.* Wiley-Interscience, New York.

Pitcher, T.J. 1986. Functions of shoaling behaviour in teleosts. In T.J. Pitcher (ed.), *The Behaviour of Teleost Fishes.* Croom Helm,

London, pp. 294–337.

Pitcher, T.J. and A.E. Magurran. 1983. Shoal size, patch profitability and information exchange in foraging goldfish. *Anim. Behav.* 31: 546–555.

Puckett, K.J. and L.M. Dill. 1985. The energetics of feeding territoriality in juvenile coho salmon (*Oncorhynchus kisutch*). *Behaviour* 92: 97–111.

Pulliam, H.R. and T. Caraco. 1984. Living in groups: is there an optimal group size? In J.R. Krebs and N.B. Davies (eds.), *Behavioural Ecology*. Second edition. Blackwell, Oxford, pp. 122–147.

Pulliam, H.R. and G.C. Millikan. 1982. Social organization in the nonreproductive season. In D.S. Farner, J.R. King, and K.C. Parkes (eds.), *Avian Biology*. Vol. 6. Academic Press, New York, pp. 169–197.

Pyke, G.H. 1984. Optimal foraging theory: a critical review. *Ann. Rev. Ecol. Syst.* 15: 523–75.

Pyke, G.H., H.R. Pulliam, and E.L. Charnov. 1977. Optimal foraging: a selective review of theory and tests. *Quart. Rev. Biology* 52: 137–154.

Rausher, M. 1979. Egg recognition: its advantage to a butterfly. *Anim. Behav.* 27: 1034–1040.

Real, L. (ed.). 1983. *Pollination Biology*. Academic Press, New York.

Real, L. and T. Caraco. 1986. Risk and foraging in stochastic environments. *Ann. Rev. Ecol. Syst.* 17: 371–390.

Robertson, R.J. and G.C. Biermann. 1979. Parental investment strategies determined by expected benefits. *Z. Tierpsychol.* 50: 124–128.

Roitberg, B.D. and R.J. Prokopy. 1983. Host deprivation influence on response of *Rhagoletis pomonella* to its oviposition deterring pheromone. *Physiol. Ent.* 8: 69–72.

Root, R.B. and P.M. Kareiva. 1984. The search for resources by cabbage butterflies (*Pieris rapae*): Ecological consequences and adaptive significance of Markovian movements in a patchy environment. *Ecology* 65: 147–165.

Rubenstein, D.I. 1982. Risk, uncertainty and evolutionary strategies. In King's College Sociobiology Group (eds.), *Current Problems in Sociobiology*. Cambridge Univ. Press, Cambridge, pp. 91–111.

Schaffer, W.M. 1983. The application of optimal control theory to the general life history problem. *Amer. Nat.* 121: 418–431.

Schaller, G.B. 1972. *The Serengeti Lion*. Univ. of Chicago Press, Chicago, IL.

Schmidt-Nielsen, K. 1984. *Scaling. Why is Animal Size So Important?* Cambridge Univ. Press, New York.

Schoemaker, P.J.H. 1982. The expected utility model: its variants, purposes, evidence and limitations. *J. Econ. Lit.* 20: 529–563.

Schoener, T.W. 1971. Theory of feeding strategies. *Ann. Rev. Ecol.*

*Syst.* 2: 369–404.

Sibly, R.M. 1983. Optimal group size is unstable. *Anim. Behav.* 31: 947–948.

Skinner, S.W. 1985. Clutch size as an optimal foraging problem for insects. *Behav. Ecol. Sociobiol.* 17: 231–238.

Slagsvold, T. 1984. Clutch size variation of birds in relation to nest predation: on the cost of reproduction. *J. Anim. Ecol.* 53: 945–953.

Slobodchikoff, C.N. 1984. Resources and the evolution of social behavior. In P.W. Price, C.N. Slobodchikoff, and W.S. Gaud (eds.), *A New Ecology*. Wiley-Interscience, New York, pp. 227–251.

Solomon, D.J. 1981. Migration and dispersion of juvenile brown and sea trout. In E.L. Brannon and E.O. Salo (eds.), *Salmon and Trout Migratory Behavior Symposium*. School of Fisheries, Univ. of Wash., Seattle, WA, pp. 136–145.

Southern, H.N. 1970. The natural control of a population of Tawny owls (*Strix aluco*). *J. Zool., Lond.* 162: 197–285.

Stephens, D.W. 1981. The logic of risk-sensitive foraging preferences. *Anim. Behav.* 29: 628–629.

Stephens, D.W. 1987. On economically tracking a variable environment. *Theor. Pop. Biol.* 32: 15–25.

Stephens, D.W. and E.L. Charnov. 1982. Optimal foraging: some simple stochastic models. *Behav. Ecol. Sociobiol.* 10: 251–263.

Stephens, D.W. and J.R. Krebs. 1986. *Foraging Theory*. Princeton Univ. Press, Princeton, NJ.

Stephens, D.W., J.F. Lynch, A.E. Sorensen, and C. Gordon. 1986. Preference and profitability: theory and experiment. *Amer. Nat.* 127: 533–553.

Stich, H.B. and W. Lampert. 1981. Predator evasion as an explanation of diurnal vertical migration by zooplankton. *Nature* 293: 396–398.

Templeton, A.R. and L.R. Lawlor. 1981. The fallacy of the averages in ecological optimization theory. *Amer. Nat.* 117: 390–393.

Tuljapurkar, S.D. 1982. Population dynamics in variable environments. III. Evolutionary dynamics of *r*-selection. *Theor. Pop. Biol.* 21: 141–165.

Turelli, M., J.H. Gillespie, and T.W. Schoener. 1982. The fallacy of the fallacy of the averages in ecological optimization theory. *Amer. Nat.* 119: 879–884.

van Alphen, J.J.M. 1980. Aspects of the foraging behaviour of *Tetrastichus asparagi* Crawford and *Tetrastichus spec.* (Eulophidae), gregarious egg parasitoids of the asparagus beetles *Crioceris asparagi* L. and *C. duodecimpunctata* L. (Chrysomelidae). 1. Host-species selection, host-stage selection and host discrimination. *Nether. J. Zool.* 30: 307–325.

van Alphen, J.J.M. and H.W. Nell. 1982. Superparasitism and host discrimination by *Asobara tabida nees* (Braconidae: alysiinea) a larval parasitoid of drosophilidae. *Nether. J. Zool.* 32: 232–260.

van Lenteren, J.C., K. Bakker, and J.J.M. van Alphen. 1978. How to analyse host discrimination. *Ecol. Ent.* 3: 71–75.

Vinyard, G.L. and W.J. O'Brien. 1976. Effects of light and turbidity on the reactive distance of bluegill (*Lepomis macrochirus*). *J. Fish. Res. Board Can.* 33: 2845–2849.

Vlymen, W.J. 1970. Energy expenditure of swimming copepods. *Limn. Oceanogr.* 15: 348–356.

von Neumann, J. and O. Morgenstern. 1947. *Theory of Games and Economic Behavior.* Second edition. Princeton Univ. Press, Princeton, NJ.

Waage, J.K. and J.L. Lane. 1984. The reproductive strategy of a parasitic wasp. II. Sex allocation and local mate competition in *Trichogramma evanescens. J. Anim. Ecol.* 53: 417–426.

Walters, C.J. 1986. *Adaptive Management of Renewable Resources.* Macmillan, New York.

Weis, A.E., P.W. Price and M. Lynch. 1983. Selective pressures on clutch size in the gall maker *Asteromyia carbonifera. Ecology* 64: 688–695.

Werner, E.E. 1977. Species packing and niche complementarity in three sunfishes. *Amer. Nat.* 111: 553–578.

Werner, E.E. and J.F. Gilliam. 1984. The ontogenetic niche and species interactions in size-structured populations. *Ann. Rev. Ecol. Syst.* 15: 393–425.

Werner, E.E. and G.G. Mittelbach. 1981. Optimal foraging: field tests of diet choice and habitat switching. *Amer. Zool.* 21: 813–829.

Werner, E.E., G.G. Mittelbach, D.J. Hall, and J.F. Gilliam. 1983a. Experimental tests of optimal habitat use in fish: the role of relative habitat profitability. *Ecology* 64: 1525–1539.

Werner, E.E., J.F. Gilliam, D.J. Hall, and G.G. Mittelbach. 1983b. An experimental test of the effects of predation risk on habitat use in fish. *Ecology* 64: 1540–1548.

Whitham, T.G. 1980. The theory of habitat selection: examined and extended using *Pemphigus* aphids. *Amer. Nat.* 115: 449–466.

Winkler, D.W. 1985. Factors determining a clutch size reduction in California gulls (*Larus californicus*): a multi-hypothesis approach. *Evolution* 39: 667–677.

Wise, D.H. 1975. Food limitation of the spider *Linyphia marginata*: experimental field studies. *Ecology* 56: 637–646.

Wise, D.H. 1979. Effects of an experimental increase in prey abundance upon the reproductive rates of two orb-weaving spider species (*Araneae: Araneidae*). *Oecologia* 41: 289–300.

Wise, D.H. 1983. Competitive mechanisms in a food-limited species: relative importance of interference and exploitative interactions among labyrinth spiders (*Araneae: Araneidae*). *Oecologia* 58: 1–9.

Wurtsbaugh, W. and H. Li. 1985. Diel migrations of a zooplanktivorous fish (*Menidia beryllina*) in relation to the distribution of its prey in a large eutrophic lake. *Limn. Oceanogr.* 30: 565–576.

Wynne-Edwards, V.C. 1962. *Animal Dispersion in Relation to Social Behaviour.* Oliver and Boyd, London.

Zahavi, A. 1986. Reliability in signalling motivation. *Behav. Brain Sciences* 9: 741–742.

Zaret, T.M. 1980. *Predation and Freshwater Communities.* Yale Univ. Press, New Haven, CT.

Zaret, T.M. and J.S. Suffern. 1976. Vertical migration in zooplankton as a predator avoidance mechanism. *Limn. Oceanogr.* 21: 804–813.

# Author Index

# Subject Index